THE FRACTAL RAINBOW
(Beyond Our Universe)

David Piñana (Mar.2017)

I0488789

To my niece Sara, that surely someday in the future will assume the Emergent-Scale-Fractal Universe as something completely usual and normal.

"Books on physics are full of complicated mathematical formulas, but thought and ideas are the beginning of every physical theory."

- Albert Einstein

COPYRIGHT

Author: David Piñana
Editor. m-dimension
m-dimension@hotmail.com
Copyright © 2015 David Piñana
V 3.0 (Mar.2017) B&W Edition
Previous editions:
V 1.1 (Dec.2015)
V 1.2 (Mar.2016)
V 2.0 (Dec.2017)
ISBN-13: 978-1519376145
ISBN-10: 1519376146

AUTHOR

David Piñana born 1958 in Spain and studied Industrial Engineering in the University of ETSEI Barcelona (1983). He has been working in the Management & Consultancy Industry during many years, and in 2012 he founded his own International Consultancy to provide funding for Energy & Environment projects over the world.

The author has been following up during many years the last advances on cosmology and quantum physic, and in 2012 wrote his first article as a disclosure text to explain the Universe scales and related physics concepts to the general public:

"The ¨Matryoshka-verses¨: The scale relativity of the Universe" *(David Piñana, October 2012).*

Afterwards he wrote the second article, to provide further ideas and arguments, and also to mention related current creditworthy studies and opinions that support the proposal and its contents (even only in some of its parts):

"Scale Landscapes (Relativity) of the Universe" *(David Piñana, October 2015).*

Based on these two articles, the author prepared the present book, through improving them and with the addition of several chapters and annexes (references by other scientists) and including comparisons between the proposal of this book and other related proposals.

TABLE OF CONTENTS

COPYRIGHT   2
AUTHOR   3

PROLOGUE   9

1. INTRODUCTION 13

PART I 15

2. THE FRACTAL RAINBOW UNIVERSE 17

THE MATRYOSHKA TRAVELS   17
THE SCALES OF THE UNIVERSE   19
UNIVERSES WITHIN OTHER SCALES   22
OUR UNIVERSE IS A VIRTUAL MODEL   26
THE LAWS THROUGH THE SCALES   28
THE LIMITS OF OUR UNIVERSE   31

3. SEARCHING FOR EVIDENCES   37

ELECTROMAGNETIC WAVES   38
GRAVITATIONAL WAVES   40
EXTERNAL GRAVITATIONAL FORCES   43
MICROWAVES BACKGROUND RADIATION   43
UNIVERSES WITHIN THE PLANCK VOLUME 44
OTHER UNKNOWN FORCE FIELDS   46

PART II 49

4. UNIVERSE LANDSCAPES   53

5. COSMIC LANDSCAPE   57

ENERGY-MATTER OF OUR UNIVERSE   59
DARK ENERGY   62
DARK MATTER (& OTHER THEORIES)   64

6. PLANCK LANDSCAPE   67

CASUAL DYNAMICAL TRIANGULATION   68
SPACE AND VACUUM 69
ENERGY & SIZE SCALE RELATIONSHIP   73
DSR THEORY (HIGH ENERGY SCALE)   75
UNCERTAINTY PRINCIPLE   76
THE NUCLEAR WAVES   79

7. EMERGENT CONCEPTS & LAWS   83

EMERGENT INTERACTION FIELDS   85
TIME CONCEPT   91
TRAVEL IN TIME   94

8. EVENT HORIZONS   95

HOLOGRAM THEORY 101

9. FRACTALS & SCALE RELATIVITY   105

FRACTAL COSMOLOGY   107
SCALE RELATIVITY THEORY   109

10. OUR UNIVERS CONSTANTS   113

EDDINGTON´S UNCERTAINTY CONSTANT 115

11. GÖDEL THEOREME VS TOE   117

PART III 121

ANNEX 1: MOND THEORY 123

MOND BASIC PRINCIPLES 123
OUTSTANDING PROBLEMS FOR MOND 126
TENSOR-VECTOR-SCALAR GRAVITY (TEVES) 127
MODIFIED GRAVITY (MOG) THEORY 128
MOND THEORIES vs FRACTAL RAINBOW 132

ANNEX 2: EMERGENCE THEORY 135

EMERGENCE CONCEPT 135
STRONG AND WEAK EMERGENCE 136
EMERGENT PHYSIC SAMPLES 139
EMERGENCE SUMMARY 140
EMERGENCE vs FRACTAL RAINBOW 141
CONCLUSIONS 144

ANNEX 3: FRACTAL THEORY 147

FRACTAL CONCEPT 147
FRACTAL COSMOLOGY 151
FRACTAL THEORY & FRACTAL RAINBOW 152
OTHER RELATED ARTICLES OVER FRACTALS: 154

ANNEX 4: BRANE THEORY 157

BRANE CONCEPT 157
BRANE COSMOLOGY 159
BRANE THEORY IN FRACTAL RAINBOW 163

ANNEX 5: SCALE RELATIVITY THEORY 165

BASIC PRINCIPLES 165
THE PRINCIPLE OF RELATIVITY 167
THE PRINCIPLE OF SCALE RELATIVITY 168
SCALE RELATIVITY & GENERAL RELATIVITY 169
FRACTAL SPACE-TIME 171

MIN & MAX INVARIANT SCALES 172

SR VS OTHER RELATIVITY THEORIES 173

SCALE RELATIVITY VS QUANTUM MECHANICS 178

SCALE RELATIVITY AND OTHER TOE APPROACHES 178

SCALE RELATIVITY VS FRACTAL RAINBOW 181

ANNEX 6: GR + QM + TOE THEORIES 185

CLASSICAL MECHANICS 187

RELATIVISTIC MECHANICS 187

QUANTUM MECHANICS 188

QUANTUM FIELD THEORY 188

STANDARD MODEL 189

TOE 190

12. EPILOGUE 193

DEFINITIONS 197

BIBLIOGRAPHY 198

ARTICLES 199

GRATEFULNESSES 200

LIST OF FIGURES 201

BACK COVER 202

PROLOGUE

Currently we assume as normal and obvious that **we are living in a big sphere (ball) that is floating in the space** and orbiting around the Sun, which, in turn, is also floating in the same space, like other stars and galaxies we know.

We also find it normal that **100 million years ago there were huge dragons (dinosaurs)**, and some of them flew and others lived in the sea.

And we also find normal that, with an object no bigger than a pack of cigarettes, **we can be talking from anywhere, on-line, with someone in another continent**.

But only **600 years ago, we were treated as witches or frauds** if we, simply, make a comment about one of them.

How can surprises us science and technology over the coming years. (50-100-1.000 years)?

Egyptians and Greeks already knew that the earth was round, and that possibly it orbited around the sun. Pythagoras (VI century BC), Herodotus (V BC) Aristotle (IV BC), Archimedes (III BC), Ptolemy (II BC), etc., they were the first scientists of humanity (Post-Flood).

But it was not until Galileo and Kepler (XVI), and Newton (XVII), when **the laws of Universal Gravitation were established** (distinguishing between astrology and astronomy). Laws that were further improved by Einstein (XX) with the theories of special and general relativity. And now they would require to be reviewed again.

Moreover, Volta and Ampere (XVIII), and Ohm, Tesla, Faraday and Maxwell (XIX) developed the **Electromagnetic laws.** While Boyle (XVII), Mendeleev (XIX), Bohr and Kelvin (XX), **separated**

alchemy of chemistry and thermodynamics, and in the twentieth century, Einstein, Planck, Shrodinger & Heilselberg unraveled **the mysteries of Quantum Physics**.

Currently **physicists are obsessed** (like also was Einstein) to obtain a universal theory that encompasses everything (**TOE: Theory of Everything**). But is this possible?, Can be a unified theory that parameterize all the laws of the universe?. It should mean that we would know everything, and that we could explain all phenomena in the universe.

The search of TOE has become the philosopher's stone of the XXI century. To this objective we have spent much of the efforts and research funding. String theory (Superstring or M-Theory), has been a major focus of attention by the academic community. But **after more than 25 years of study, it has not been able to reach any definitive result**. Something is wrong, and we need to find out what.

It has been proposed that the solution may be to **better understand the essence of space-time**. And, yes, I am sure that this is so (consider space-time as fractal concepts can be a solution). But I also think that it is not all that fails. And also we must **consider the concept of "Spatial Scale" as a variable** to be included within the physic theories (Scale Relativity).

Moreover, **in this book we will show how far we are to know everything**, and also to be able to have a TOE. It is most likely that this objective will be not feasible, and that, at best, we should limit ourselves to understand only the universe closer to us (from a scale viewpoint). Although, this scale spectrum, can go expanding as we progress technologically and scientifically.

I am almost convinced that **one day we will travel faster than current light speed (300.000 km/s)**, we will **contact with beings from other star systems**, we will be able to **travel to other 3D parallel universes**, and, it is even possible, that we will **communicate with beings from other dimensional (size) scales**. Possibly, we also will be able to **teleport**, to communicate by **telepathy**, and to be able to **understand the phenomena of the apparitions** (UFOs, ghosts, spirits, "ouija",…). And, possibly, we also will know **who built pyramids in Giza**, who were the **first habitants of the city of Tiwanaku** (on lake Titicaca, Bolivia), and what lies **beneath the Temple of Solomon in Jerusalem**.

But I am also sure that **we will not be able ever to travel to other spatial (size) scales**, whether at higher or lower scalar sizes (we shall "never" be able to do matryoshkans travels). And, I fear that, we also we will not be able ever to travel **into the past**. At least in a way that we can interact with it (what we usually understand by traveling physically to a place), although we can "visualize" what happens there. And no more due to a technology limitation or scientific knowledge, but simply because it is not possible from a logical and conceptually point of view. In other words, **these last proposals are absurd.**

In Abril.2012 I was with an agro-ecological friend that is very proud of his relationship with nature. That day we were having the lunch in his cottage. I was surprised when he asked to me what were the "strings". He had read something about it in a magazine and wanted to know what they were. **This was the beginning of this book.**

Initially, I just wanted to shape (in one article) for an **explanation of the different scales of the universe**, and the different concepts and laws prevailing in each space scale, and to suggest that possibly these scales could be expanded in the future, when technology and our knowledge improves.

But to go deeper, I feel more and more intrigued to see how **such trivial concepts as matter, energy, space and time**, they **were getting a very different aspect** that we have of them intuitively.

As we shall read in the book, **what we could consider matter,** it is concentrated in certain particles (bosons) with a volume of 10 e 24 times smaller than an atom. So the matter **consists mainly of empty space**. And solid, liquid and gaseous states are due to the electromagnetic charges of atoms.

But the vacuum, in turn, is like another type of substance or state of matter-energy. Perhaps **we can consider matter-energy and empty space (vacuum) as different types of stages (or phases) of our own reference scale.**

On the other hand, **time** would be only a way of visualizing (measuring) the movement, or the change of events in motion (entropy). If nothing in the universe would move (change) then it would not need the time. **In a static universe there would be no time.**

Energy, matter, space (vacuum) and time are emergent concepts of our own reference scale. For each size scale (scale landscape) there will be other concepts, and other laws that govern them.

To assimilate and to accept the message of the last few paragraphs represents a complete **change of vision of Our Universe**, and it offers a new perspective for understanding and modeling it. Opening new expectations and paths to study.

It is logical that **science is resistant to change**, as also to new proposals not properly credited (proven or experienced). So we have progressed throughout history. **But neither it is so good to be contrary and too strict with changes**. It could be absurd to invent new strange concepts (as Dark Matter could be), just to not accept that, possibly, the current laws (Newton and Einstein) are not valid for certain conditions (very large scale dimensions). At least the same priority should be given to both options, which is not the case today, where Dark Matter is the priority alternative, leaving the other virtually forgotten.

We have to accept that **we are at the very beginning to understand the whole Universe**, and that we have to be prepared for future great discoveries, and they could be almost unexpected now for us

I hope that this book could provide a bit in this ambitious project, and that the **reader enjoy while reading it,** as much as I have enjoyed in their preparation.

1. INTRODUCTION

The current mainstream in cosmological physics proposes a unique universe that started with the Big-bang (approx. 13,700 million years ago), and it continues expanding "isotropically". This "mainstream" universe seams to be limited by both scale spectra: in its upper level (approx. 10 e +27 m.) by Our Universe external "shape boundary" (or, at least, by the Observable Universe one), and on the other hand, on its lower level (approx. 10 e -35 m.), by the Planck scale/size. **And we (humans) are in the middle, at intermediate levels/spectra between these scale/ size limits.**

This book intends to show a **new conception of the Universe: The scale relativity**. This book proposes that the universe is composed of many more scale spectra (upper and lower) than the currently recognized limits (possibly over 10 e +1000 m and below 10 e -1000 m). And for every spectra there would be different physical concepts and laws (emergents), although they could be linked by common underlying laws and concepts. And we (humans) are not in the middle. **We are just in a random level within this broad spectrum scale.**

This proposal (if it is true and it can be proved) **might be a very important advance** in explaining certain physical concepts that are currently not entirely clear (Dark Energy and Matter, Uncertainty Principle, particle/wave dual nature, etc.).

Humans, as always, make the mistake of believing that **we are (in) the center of the universe**. And we always try to understand the universe from the Ptolemaic point of view. But time after time, we have had to admit that this is not so (Copernicus,…).

We are now making the same mistake, and **we believe that we are in the middle of the scale (size) spectrum of the universe.** And we try to understand the universe from what we know about our own scale (size) spectrum. Trying to extrapolate our own concepts of scale (size) to other scale spectra. Probably, there would apply other concepts and other laws, unknown to us (emergent concepts and laws).

Surely, if we are able to break these pre-established schemes, **we will open new horizons**, allowing us to understand physical concepts that currently we fail to understand well: Dark Energy and Matter, the Uncertainty Principle, the dual nature of the particles (particle/wave), etc.

The universe could be composed of many more scale/size spectra (upper and lower) than the currently recognized limits (possibly over 10 e + 1000 m and below 10 e - 1000 m). And **we humans aren´t in the middle**. We are just in a random level within this broad spectrum scale/size *(See Fig.3)*.

As Lee Smolin says, to any major new physical theory (such as Newton and Einstein), assuming that, once it is proposed and accepted, they would clear other big uncertainties to date. As in the game of dominoes (where after falling a first tab, then fall all the others), so it happened with many concepts when considering the theories of Newton and Einstein. And it can happen the same, if we can demonstrate and accept the approach of the **Scale Relativity of the Universe**, proposed in this book.

It is not the objective of this book to give a clear mathematical / theoretical demonstration or experimental verification concerning its proposal. This book just try to propose a *framework*, and to show several collateral and related studies and theories that fit within this framework. So **we could classify it better within the Scientific Philosophy than within the Cosmological Physics**. To get a final **mathematical/theoretical demonstration** will be the following objective to follow up, that will be not easy. Furthermore (in parallel), we have to look also for **experimental verifications**, that surely will also be very difficult. Multidisciplinary team work and high investment will be necessary to get them !

PART I

This book part contain mainly the reviewed first article:**"The ¨Ma-tryoshka-verses¨: The scale relativity of the Universe"** *(David Piñana, October 2012).*

The main objective of this article was to show to main public a different point of view of the Universe: from its different space scales.

It shows the Universe as a 3D-Fractal Rainbow (space scales spectra) where in every spectra (landscapes) we could have different concepts (atoms, stars,…) and laws (Newton, Quantum,…).

But also that these spectra (landscapes) could be infinite, and that current "mainstream" boundaries of Our Universe, could be enlarged in the future.

And, mainly, that we (humans) are not just in the middle of them. We are just in a random spectra (landscape).

2. THE FRACTAL RAINBOW UNIVERSE

THE MATRYOSHKA TRAVELS

Everyone knows the movie **"Fantastic Voyage"**, (based on the novel by Isaac Asimov, 1966). Where it is reduced the size of a submarine (Proteus), and also its crews, to be introduced into the blood circulatory system of a person, with the objective of heal the person of the blood clot he has in the brain.

ARGUMENT:

The book "Fantastic Voyage" (Richard Fleischer, 1966) begins with the crash of one of the most important members of the medical field. Due to this cause, a clot forms in his brain that must be removed to survive, but this operation can not be performed from the outside.
To do it, they decide to introduce a miniature submarine inside the body. The submarine can only last 60 minutes before returning to normal. To carry out this mission decide to choose their best experts, and they will be who will go inside the submarine, to make the medical intervention.

The film has good scientific basis, as always Isaac Asimov do, but also it has some important conceptual errors, both, from the point of view of medicine / biology, and physics / mechanics: Circulation of the submarine in the blood (difference between blood and water density, visibility,..).

But, possibly, the most obvious paradox, and possibly the less understood by the reader / viewer, is that, **by the fact of reducing a person's body, it is required also to reduce the component parts in the same way (cells, molecules, atoms, protons, ...).** How it is possible that equivalent elements (Ex: water molecule) can coexist in the same system with different dimensions (up to 10 e 5: 100,000 times)?. How the tiny intruder would fit within the universe that has been implanted?. What laws of physics / chemical / biological would be applicable (micro, macro, or a combination of both), and How they interact one with each other?

> *Clearly this is not "conceptually" possible, so that **these trips between different sizes (Matrioshkanos) are absurd.***

In physics and engineering also there is the **Scale Models Theory** (to test prototypes at small scales), to simulate the behavior of certain variables parameters, to be extrapolated to the normal scale, and **to establish the corresponding model-prototype similarity** (Geometric, Kinematics and Dynamics): **between real and scale models**.

Even with their errors and paradoxes, this book / movie gives us an idea of what would be to travel between different scales (powers of 10) of the space dimensions (XYZ). This is what we might call **"Matryoshka Travels"** in reference to the famous "Russian Dolls" called Matryoshka.

Likewise, it also evidence us about the difficulty and strangeness of these trips to smaller scales (Negative Powers: cells, molecules,...), where we cannot travel otherwise than by making us so small as the size of the scale where we want to travel. For this film the reduction is 100,000 times, but if we would like to move to the scale of the electrons, we would have to reduce our size 1,000,000,000, 000,000,000 times (10 e -18).

However, for traveling to larger scales (Positive Powers: galaxy, universe,...) we always imagine it without having to change any scale or dimension. It is assumed that a spacecraft will be used (interstellar or inter-galactic) and we will move at high speeds (maximum speed of light). But we never imagine to travel there making us (and the spacecraft) larger: increasing our size as well as we did in the case of traveling to the small.

If we make this effort, and we made this change of mind (although the concept itself is not scientifically feasible) we could visualize what would be the new concept of the Global Universe: **nD-Fractal Rainbow** (or **"Matryoshka" Dolls).** The nD means the dimension of the space (we will understand it latter on). For Our Universe n=3 (3D).

Usually, when we conceive beings from other worlds (other stars, galaxies or universes), we imagine them in a human scale. Or, what is the same, coming from the same spatial scale that we (humans) are. But we never imagine them coming from other different spatial scale (larger or smaller) to ours. In the case that the scale would be very different, it would impede to us to view and realize them (because they could be too much large or small).

THE SCALES OF THE UNIVERSE

It is normal to visualize the different scales of the known universe in powers of 10 and the meter as the base of measure (see table below):

• **Negative Powers**: For smaller scales (till 10 e -35 meters for the Planck length).
• **Positive Powers**: For larger scales (till 10 e+35 meters for "multiverses").

See link: *http://www.microsiervos.com/archivo/ciencia/escala-universo-interactiva.html*

This show us a Global Universe compound or divided into different zones (spectra, ranges, ...) that form the different **Space Scales of the Universe**.

Based on a scale of one "meter" (this is the power 10 e 0), as we increase the exponent of 10 positive (1,2,... , n), we will show larger sizes (10 e 3 is 1 km , 10 e 9 is a million km, and 10 e 16 is a light-year, the distance light travels in one year).

Some dimensions of reference (m e +10):

Concept	Exp	Discovery
Man	0	
Earth	7	II BC (Eratosthenes measured diameter)
Sun	9	
Solar System	13	XVI (Galileo & Kepler)
Galaxy	21	XVIII (Herschel)
Universe **(The farthest detected)**	27	XX (Big-Bang of George Gamow)
Multiverse	35 (¿)	XXI (to prove and see)

Instead, if we make powers 10 negative, we will define measures under (10 e -3 is a millimeter, 10 e -9 is a nanometer and 10 e -35 is the Planck Unit).

Some dimensions of reference (m e -10):

Concept	Exp	Discovery
Man	0	
Cell / red blood cell	-6/7	XVII (Robert Hooke)
DNA	-9	1953 (Francis Compton-James Dewey)
Water Molecule	-10	XIX (Amadeo Avogadro)
Protón **(The smallest detected)**	-15	1919 (Ernest Rutherford)
Electrón/ Quark	-18	1897 (JJ Thomson) &1950 (M.Gell-Mann)
Neutrino	-24	1930 (Wolfgang Pauli)
String	-35	XXI (to prove and see)

The knowledge of the composition and the laws governing the different levels of scales (both, positive and negative) has been discovered through the ages (Periods of History) as scientific and technological advances. The advances in large (**classical and relativistic physics**) have been produced earlier by the ease of being observed (during the last five centuries), while the small (**quantum physics**) have occurred mostly during the last century.

In each of these zones (levels, spectra,...) are supposed to act or apply (in a greater or lesser extent) some physical laws. These laws that would make that the elements and form structures coexist with a certain harmony and logic, as it does in our own zone/spectra of **Our Universe**.

We could say that **Our Universe** covers scales from 10 e -35 (**Planck scale**, the smallest allowed by the current physics models due to the emergence of quantum gravity effects) to 10 e +27 meters (which is the size of **Our Universe**).

The visual (or photographic) observation of these elements (both positive and negative scales) requires optical magnification systems or other wave detection systems:

- **Telescope** (optical, x-ray, infrared,...) for the observation of large bodies to long distances (electromagnetic waves). The most distant objects that have been detected are 13,000 million light years (10 e+25 meters).
- **Microscopes** (optical, electronic, nuclear, ...) for the observation of small bodies at very short distances._Smaller objects that have been detected are 10 e-15 meters.

In both cases, further the ability to increase (produced by the lens), one of the factors that hinder their vision is the lack of light (or other waves) as we increase the scale of vision.

In both cases, it is required high power glasses, and increased sophistication of technology or a long exposure to capture light (electromagnetic wave) needs to be viewed or recorded. And even in some cases requires the use of other wave detection systems (Telescope: X-ray, infrared,... and Microscope: electrons,...) to better detect these bodies.

It is like moving away within this **3D-Fractal Rainbow** (both positive and negative), from our reference range/strip, we remained without the stimulus (waves) that we (humans) require to observe or record a form or body. It's the same that occur when we move in one of the spatial dimensions (XYZ) , that it is becoming smaller and blurred the body or form that we move away.

From our point of observation, we are in the middle of our spectrum or band of the **Universe Global** (**Our Universe**). As further away we move from the scale of the bodies and entities, we find it more difficult to observe and detect them. And whose boundaries have been extended through time, as that science and technology have progressed.

In an interview with **Eduardo Punset (TV2, Spain)** the renowned astronomer and physicist **Stuart Clark** (2012), this last commented: *"... Our specie (humans) is halfway between the largest structures and the smallest of the Universe, and we are very fortunate because it serves as a platform to observe the universe in all its dimensions ..."*. According to the theory proposed in this book should be this statement : *"... Our species is halfway between the largest structures and the smallest of the Known Universe because from the reference spectrum scale where we are, it is what our technology and expertise allows us to see and understand. But it is very possible that in the future this range will be widened in both directions, and it will allow us to see and understand the universe in broader dimensions... "*

UNIVERSES WITHIN OTHER SCALES

For centuries humanity has been assumed that the universe was the Earth (flat or not) and Sky (with stars and planets) going around. The Earth was the center of the Universe over which turned the other bodies. And our scale

Something similar happens us now, where we assume that the area (level, spectra, ...), where we are within the **Global Universe,** is in the middle. But there could be other universes in other areas within this **nD-Fractal Rainbow** that will have their own physical laws, and other types of forms or bodies, and even, why not, other living things. And our scale spectra does not necessarily have to be at the middle.

If we can imagine (as an allegory) that an electron is the equivalent to the Earth revolving around the nucleus of an atom, like do the Earth around the Sun. How we would see our universe if we lived in this electron? We would imagine that we are part of another body or entity? As much as we travel with a spacecraft of this scale (10 e -20), it would be very difficult for us to observe the entity that we belong to. And also it would be very difficult to receive the waves to display or capture the required information.

Possibly at this scale (**negative powers**), there are other waves very much smaller (both in frequency and scope) and undetectable to us nowadays. These waves are what we would be able to detect if we were living in an electron, and of course, the waves we mean by light (electromagnetic optical spectrum) would be undetectable for us and the instruments that we would have in these dimensions (10 e -20 m).

On this scale of negative powers prevail the nuclear fields/interactions (weak and strong), with the electromagnetic field, while the gravitational field has less influence. And it s behavior can be explained by of **Quantum** physics model (and mathematical model of **Neumann**).

In these scales, it is an Universe where coexist entities such as photons, bosons, graviton, neutrino, positron, etc.. And there is a duality between waves and particles. And the matter, as we know, becomes meaningless. And the **Uncertainty Principle** (Heisenberg) is its essence more representative (at least, observed from our scale).

REFERENCES OF SCALE
OF OUR UNIVERSE

Dimension of	Exp of 10 (meters)	
Multiverses	35	
Universe	27	
		(The farthest detected)
Galaxies	21	
Light year	16	
Solar System	13	10.000.000.000.000 meters
Sun	9	
Tierra	7	10.000.000 metros
Radio Wave	0	1 meter
Microwave	-2	
Wave Infrared	-5	
Light wave Violet-Red	-6/7	Red Blood Cell
Ultraviolet wave	-8	
Virus / DNA	-9	
Water Molecule (H2O)	-10	0.0000000001 meters
Hydrogen atom (H)	-11	
Wave X-Ray/Gamma	-12	
Average Atom	-13	
Nucleus of the átomo	-14	
Proton/Neutron	-15	
		(The smallest detected)
Quark/Eletron	-18	
Neutrino	-24	
Strings (Plank Dimension)	-35	

Fig.1: References of scale of Our Universe

The new theory Causal Dynamical Triangulation (abbreviated as CDT) propose that space-time will be **two dimensional (one space and time)** near the Planck scale, and reveals a fractal structure on constant time slices. *"Where Dynamic selects a particular preferred dimension between the three possible. This dimension would be given preferential classically and then it would randomly alternating physical processes that are operating at those scales "(Carlip, 2012).*

For a civilization that, hypothetically, could be on these negative scales, **the time scale should also be very different** from our reference scale or our human scale. Possibly the emergence and end of a civilization, on a scale 10 e -20 meters, would take time proportional equivalent of our scale (For example, the human civilization: 10 e -20 x 100,000 years = 10 e -15 years = 10 e -8 seconds = 10 nanoseconds).

Likewise, we can extrapolate this experience to large scales (**positive powers**). Imagine (as an allegory) that a galaxy is a type of cell (neuron) of a Superior Entity, and that the interaction between these type of neurons (Galaxies) with other galaxies (neurons) make a thought (set of bits) of this Superior Entity.

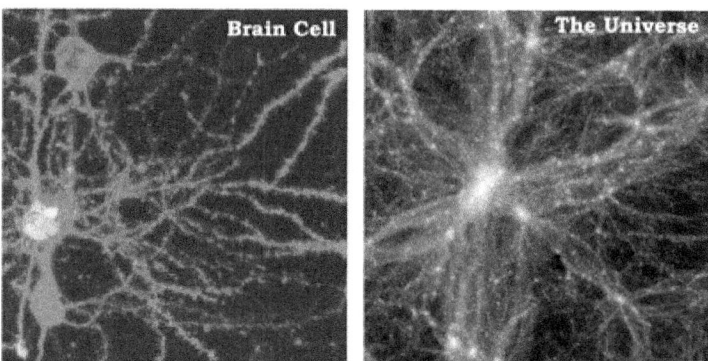

Fig.2: Images:Brain Cell and Universe Galaxies

At this scale of positive powers, prevail the **gravitational and electromagnetic** fields (forces), and the influence of **strong and weak** fields (strength) are less, or practically zero. And its functioning currently can be explained by the physical models of the Theory of Relativity (**Special and General Relativity** by Albert Einstein) and mathematical models of **Riemann.**

At these scales, there coexist entities as black holes, galaxies, nebulae, etc.. And there is a duality between matter and energy, and between space and time. And the speed of light ,as a constant and a limit of our universe, takes all its prominence. And there the **Theory of Big Bang and Isotropic Expansion** of our universe are its essence more representative (at least seen from our scale).

As we move to increasingly larger scales, and observed from our reference scale, it seems as if we need a **fifth dimension of space-time.**

For the civilization that, hypothetically, could be on these positive scales, **the time scale could also be very different** from our reference scale or our human scale. Possibly transmitting a stimulus between a neuron (Galaxy) to another neuron (Galaxy) on a scale of 10 e+20 meters will last for a proportional time equivalent to our scale (10 e+20 x 1 microsecond = 10 e+14 seconds = 10 e +7 years = 10 million years).

For an observer who was in either of these two scales (electron or macro-neuron), **His Universe** would be another. And it would cover other spectra or bands of the Global Universe.

At these spectra of the **nD-Fractal Rainbow**, as we have seen, govern other models and patterns, as well as other waves and stimuli, other fields and forces, other entities and bodies.

If we refer to the allegory of the **nD-Fractal Rainbow**, we could say that the universe of the electron would be in the **"blue spectrum",** while the neuron universe would be in the **"yellow spectrum"**, while Our Universe would be in the middle, in the **"green spectrum".**

The latter would be our band, which is easily explained by the mathematical and physical models of Euclid, Newton and Maxwell. Where everything seems normal and takes logical sense. Where our brain and our senses are comfortable, because they have evolved to survive in it: **"The Green Strip or Spectrum"**.

OUR UNIVERSE IS A VIRTUAL MODEL

As that happened to our ancestors, we are slaves of the information we receive through the senses (sight, hearing, smell, taste and touch) and we process this information according to the knowledge we have.

The universe, as we know (perceive) it, is a virtual model (an illusion) that our brain shapes from the stimuli we receive through our senses (sight, hearing, smell, taste and touch). These senses have developed and evolved through natural selection for survival, and depending on the stimuli that exist in our universe (electromagnetic waves, pressure waves,...). As well as in other animals have developed other senses for other stimuli (ultrasound waves, ...).

We could imagine entities (some of them already exist) that have organs (senses) sensitive to other stimuli such as X-rays, infrared or ultraviolet rays, radio waves, etc.

But it seems clear that the stimuli may be different in each zone or area of the M-Dimension. There will not be the same stimuli (waves,...) at quantum scales (<10 e -10) that at large scales (> 10 e +10). And they will be more different to much further scales from our universe (scales smaller than 10 and greater than 10 e +50). Although these scales are difficult to assimilate for us, and possibly they will be not feasible, according to the current scientific and physical models.

---

*If we consider that one atom measure (on average) 1 Armstrong (1 Å = 10 e -10 meters) and the nucleus is 10 e -14 meters, means that the nucleus of an atom (where there is the mass of the atom) has a diameter 10,000 times smaller than the atom itself. If we suppose that the atom had a diameter of 100 meters (a football stadium), the nucleus would measure 1 centimeter (as a button). Put in another way, in one atom fits 1,000,000,000,000 (10 e 12) nucleus. So that, **for every 1 volume of mass there are 1,000,000,000,000 of vacuum**.*

*Furthermore, if we consider that the nucleus consists of protons and neutrons, and these, in turn, are composed of quarks, and that quarks are also 1.000.000.000.000 smaller that the nucleus, then we can say that **quarks (and electron) are 10 e -24 smaller (in volumen) than one atom.** If we consider a diameter of a quark/electron = 1 cm (like a button), the atom will have a diameter of 10.000.000.000 cm (= 100.000 km !).*

*But, furthermore, we know that **quarks are not solid particles, and they are more like wave base particles** (probability wave function), and they are possibly constitute by strings, **we can realize that the concept matter (mass) is very ethereal/virtual.** And, as we will explain latter, they could be emergent concepts.*

Therefore, what we perceive or recognize as forms of **Matter** in the Universe (bodies), as a stimulus of the sense of touch, they are only forms nearly empty (with a mass per volume of 10 e +24 of vacuum), but with electrical charges (positive and negative), which are what really give us the sensation (touch) of consistency of the bodies, and the different states of matter (solid, liquid and gas). These are the electrical charges that prevent solid objects transferred. If we can neutralize the electrical charges of a solid body (a ball), this could trans-pass another solid (walls).

The **Pauli exclusion principle** *is the reason why we cannot penetrate solid objects with hands.*

And, if we consider that the **Colors** are only electromagnetic waves ranging from red (700 nm = 7 10 e -7) to violet (400 nm = 4 10 e -7). And they are produced by altering the orbits of electrons in molecules by excitation of photons colliding with them. If no photons (light) arrive to the object, waves are not issued and we do not see the body colors. Depending on the type of molecule, there are different waves giving the different colors that we can see.

All this show us the virtual universe in which we live, and we do a composition according to the stimuli we receive. It's very similar to what might be a computer program, similar to **"Second Life"** or any 3D virtual game. Where in the smallest we can find the "pixels" of light (the smaller they are, then higher is the quality definition). And for the largest, we have to define limits, establishing loops that the user do not notice them.

Extraordinary is the scene from the movie **"Level 13"** (directed by Josef Rusnak in 1999) in which the protagonist discovers that he is in a virtual computer program.

THE LAWS THROUGH THE SCALES

We are pure energy properly conjoined and harmonized by the Fundamental physical principles and laws underlying the universe. But these laws manifest differently across the different spectra of the **nD-Fractal Rainbow**. Being nD = n Dimensions, we can understand that the different scale spectra can be of different dimensions (D-Branes).

Different force known fields (weak, strong, electromagnetic and gravity) govern differently across these (scale) spectra or zones of the **nD-Fractal Rainbow**. That has forced to use different models and patterns so that we can explain and interpret them (**Classical, Quantum and Relativistic Physics, and mathematics of Euclid, Neumann and Riemann**).

We can accept or agree that the **Fundamental and Underlying Laws of the Global Universe** are (or can be) the same for all spectra (and branes), although they manifest themselves differently through them.

Thus, these manifestations of the physical laws of the different levels are not independent one of each other, but rather dependent and consistent one with each other, so that may be established models and theories that encompass several zones or spectra, as Quantum Mechanics, the Theory of Relativity, or the lately **M-Theory**.

And these areas may even be (in some bands or scales) **parallel universes** with their fields, waves, stimuli, forms, beings (living or not).
It will be the same that for a universe of **three dimensions (2 space X-Y and time) it appear a Fourth Dimension (Z):**

We could suppose that a sheet of a book could be this Flat (2D space) Universe, where possible entities could live and move across its surface XY. every sheet of the book could be a different flat universe (parallel universes), and could be as many universes as sheets has the book. The entities living and moving by one sheet will not know about the other sheets (or universes), unless there is some kind of "shortcut" ("wormholes") or strangr communication ("ghosts"?). As in the book of Edwin A.ABBOTT, "Flatland".

If we see the diagram that represent the different spectra (bands, zones, levels, ...) from the scale 10 e -1000 to 10 e +1000 in intervals of 10 e 100, we can get an idea of how big the **Global Universe** can be compared to **Our Universe** (known one), that moves only in the central strip (10 e -50 to 10 +50) where, at best, really we get to know or intuit only from 10 e -35 (strings) to 10 e +35 (multi-universes).

Clearly, these spectra are not limited just to these borders. But these bands or spectra and colors should be understood as if they change in a gradual way (like the colors of the rainbow) and **unbounded**. There may be some effects that only cover a part of a strip (air pressure waves, sea waves, ...), and other various levels (electromagnetic waves, gravitational field, ...).

If this proposal about a Global Universe that consists of different bands like a **nD-Fractal Rainbow** *("Fractal", because although in the diagram are represented in bands of 10 e10 m it is evident that these scales are logarithmic and these bands are smaller for smaller exponents)* was correct, then there could be electromagnetic waves at other scales that currently would not be detectable by our systems:

- Both, for <u>large levels</u> (eg. on scales of 10 e +100 meters), which surely would be constantly going through Our Universe, although we could not detect them (or yes we could?). Electromagnetic waves of 10 e +100 meters wavelength could be passing through Our Universe, undetectable to the tools and technologies currently available.And, why not 10 e +1000 meters?.
- As for <u>small levels</u> (eg. exp scales of 10 -100 meters), which, considering their high energy levels, might have a very short range (within their own scale).

As we have mentioned above, at these scales also there could be another type of waves and fields, which are currently completely unknown to us. And they would own to the different physical models that would govern in the **(Open) Physics System** of these scales.

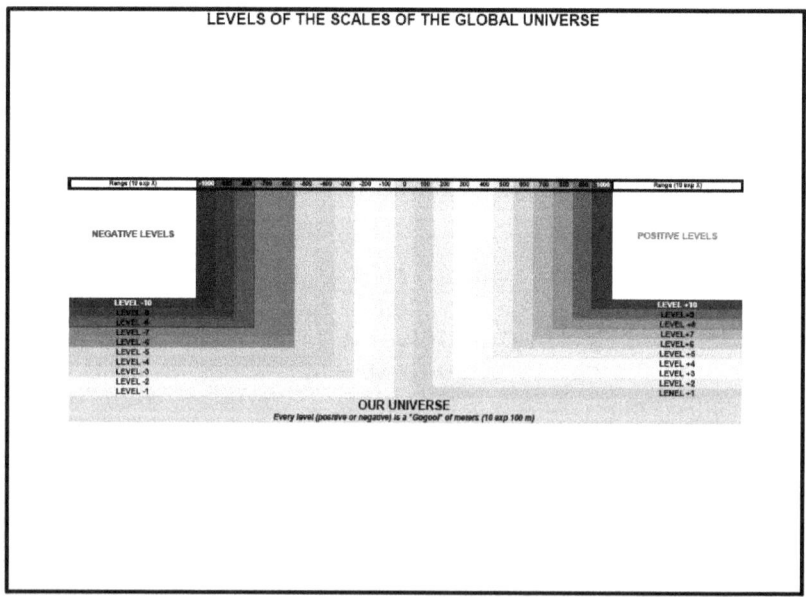

Fig.3: Levels of scales of the Global Universe

If these **Scales Levels** could be considered as **Open Physical Systems**, it would mean that there could be exchange of energy between them. Then, Our Known Universe, could absorb or lease energy from other scale levels (lower or higher) without contradicting the **First Law of Thermodynamics** *("Energy can not be created or destroyed, only transformed"...* or it **change to another level or Scale System !!!**).

THE LIMITS OF OUR UNIVERSE

To be consistent with the current existing physical models (mainstream),we have to accept that Our Universe has its limits within this **3D-Fractal Rainbow**:

- A down limit: the **Planck Scale** of about 10 e -35 meters. Any lower dimension can not be treated adequately with the current physics models due to the emergence of quantum gravity effects.
- A top limit: the size of **Our Known Universe** about 10 e +27 meters.

So, our known **Spectra or Zone of the Universe Global** has a total "scale" order of magnitude of **10 e +62**, less than a Googol (10 e 100).

According to the currently mainstream concerning the cosmology theories, **Our Universe** is considered as a closed physical system and that it is isotropically expanding. This leads to inconsistency of requiring a **Dark Energy** to explain this expansion, so as to have to accept that the **expansion rate** may be higher than the speed of light.

Possibly, if we consider **Our Universe** as an **Open Physical System** whether these inconsistencies could be solved (?).

The Hubble Space Telescope detected during 2003 and 2004 the area known as the **Hubble Ultra Deep Field (HUDF),** which displays what is believed to be the first galaxies after the Big Bang, and they are more than 13,000 million light years away. So they are the most distant objects ever observed by humans.

According to the **Theory of Isotropic Expansion of the Universe** (which expands equally at the same speed in all directions and zones), there is no center of the universe which generated the Big Bang, and there are no limits or external borders of Our Universe.

What has been seen in the HUDF can also be seen in any direction we look (in fact, have been carried pictures of other zones of the Universe with galaxies similar to the HUDF). So the **HUDF not focus toward the center of the universe or the point of the Big Bang**.

Isotropic Expansion is easily understood if we think of 4 coins (A, B, C and D) spaced 5 cm, which every 10 seconds we separated 5 cm. At the initial instant of separation between A and D would be 20 cm. But after 1 minute they would be 240 cm, while the speed of separation between them is only 30 cm per minute. If we imagine this idea considering the magnitudes of many stars and galaxies in the Universe, and the separation speeds, you can understand why it is assumed that the overall rate of expansion of the universe can exceed the speed of light .

A property of the *Isotropic Expansion* is that, from anywhere in the universe, we will see that objects (stars, galaxies,...) are distancing from us at a speed proportional to the distance they are away from us. Then if we put a limit to the expansion rate (Eg The speed of light), this implies that the expansion distance stop. If we look from Earth in all directions would have a border where the expansion would stop, and we (the Earth would be in the center). And this would be very strange and very casual.

It is interesting to understand the difference between **Our Universe** (known) and our **Observable Universe**. Because the isotropic expansion occurs at a speed exceeding the speed of light, it is assumed that **Our Universe** (known) may have a diameter of 90,000 million light years. While, such as occurred approximately 14,000 million years ago, the **Observable Universe** is a sphere with a diameter of 46,500 million light years, **within our universe**.

Furthermore, and according to current conventional physical theories, Our Universe has no center, no boundaries, no limits outdoors. It has not a particular 3D morphological form. At these scales we talk about **forms of 4D space-time**, which make them difficult to assimilate for our brain, and it could be only be modeled mathematically (Friemann Equations and Riemann "Manifolds").

Different alternatives are possible (space-time Universe Flat, Closed / Spherical or Open / Hyperbolic). And if the universe were flat (that it is considered more feasible option), the form of 4D space-time of the universe could be a **4D Möbius Toro** or **4D Klein bottle** that they are Closed Loop Systems (loop of space-time that prevents anything can come out of this "4D volume" of space-time). It is neither more nor less than what we do to limit programs 3D computer games, to never exceed their limits, and without realizing it.

One way to understand the **closed-loop systems** is our own planet (Earth). For many centuries and thousands of years humanity considered that they lived on a **Flat Earth** (infinite or not, and where sea water is poured by their limits), where the Sun, Moon, planets and stars swirled around. Now we know that it is spherical, and that if we walk straight to direction we want, after 40,000 km, we come back to the same place. This is a **closed loop of 2D** <u>curved in 3D space</u>.

The fact for which our known universe is present in such forms as complex 4D and not as a sphere or, at best, as an irregular cloud, is because it is considered that in these dimensions <u>the space-time collapses on itself due to the enormous force of gravity.</u>

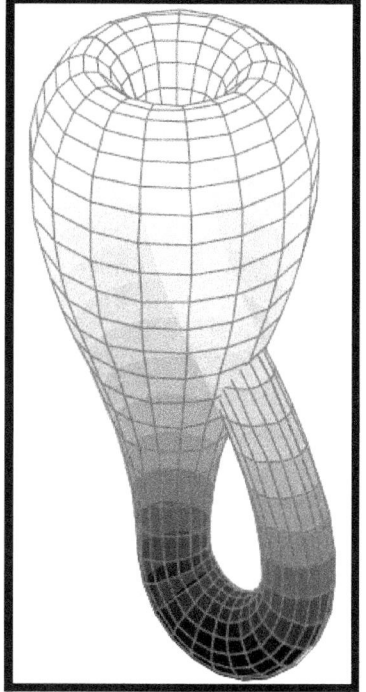

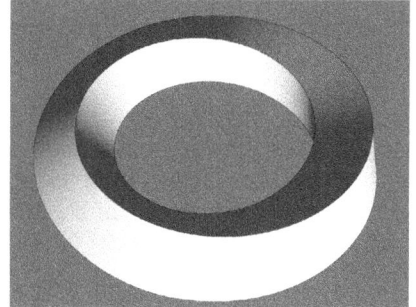

Fig.4-5: Possible 4D shapes for Our Universe

4D Möbius Toro or **4D Klein bottle.**
(3D shapes representations)

Despite all this, who knows if someday we could detect objects outside of our universe, or even to escape from this annoying **4D Closed Loop System** in which we are. **Although for viewing this Universe from outside, our brain should be able to assimilate and understand 4D forms.**

Our brain (and entire body) has evolved to survive and understand the scale range or level in which we live (between 10 e-10 and 10 e +10 meters, between the molecules and the solar system). To understand the upper and lower scales it should make an extra effort, and rely on mathematical models, which are sometimes, completely impossible to understand by our brain. Are the scales between 10 e -30 to 10 e + 30 the limits that our brain is able to assimilate?

To better understand what has been said we can take the **example of an ant**. Is its brain ready to understand that the Earth is round? And what is it understanding when a human touches it or prevents tits passage? For it this will be considered a phenomenon of nature, like an earthquake or hurricane.

It has long been considered many nebulae and galaxies visible to the naked eye as simple stars or nebulae, respectively, until the instruments allowed us to discern, and detected many galaxies. Currently it is believed that there are so many galaxies (10 e 11) and stars (10 e 11) in our galaxy (Milky Way).

Who knows if one day, we might be able to detect other objects or universes outside the boundaries or horizons of **Our Universe**. Or we detect waves or any other type of stimuli, which came from outside our universe, and out of this **Closed Labyrinth 4D**.

It would be absurd to think that the currently known limits of the Universe are their absolute limits. This would imply that we already know almost all the physical laws of the universe. But common sense tells us that it is more feasible that we are at the beginning of knowing, and that we still have much more to discover and understand.

We could consider the **dimensional network of the Global Universe** (3D space + 1 time for Our Universe, described by Einstein's Theory of General Relativity with deformations due to the Gravity and the Energy) as a net of **different N dimensions for different size scales.**

*"Why comply with the known models and patterns of **Underlying Fundamental Laws of Our Universe,** if we could understand them better if we could decode them from <u>outside its boundaries</u>?" (the common sense).*

KALUZA-KLEIN THEORY: THE EXISTENCE OF NEW SPATIAL DIMENSIONS <u>(See also ANNEX 4)</u>

*Theodor Kaluza (1919) and Oskar Klein (1926) proposed a theory that assumes the existence of <u>higher-dimensional space</u> (to the three known XYZ<u>) wound on these at very small scales</u> (Planck size). These would be **circular dimensions new and independent** directions. As if we look a hose from a distance seem to have only two dimensions, but, as we approach, we see it has another that forms its outer surface._According to **String Theory**, these extra dimensions would be 6, and these should be rolled in a **Calabi-Yau 6D** shape. This theory could be compatible with the CDT (Causal Dynamical Triangulation). For our scale, these dimensions do not affect us, and they cannot be appreciated. But for hypothetical residents within this size so small, they would take all their meaning and prominence.*

*We might ask whether the form of 4D space-time of our universe (4D Möbius Toro or 4D Klein bottle) would be just like these other **dimensions of Kaluza-Klein rolled but for large dimensions** (4D Loop System closed). So if we went out of Our Universe, certain dimensions that we detect inside would cease to have effect. It is as we assert that the **dimensions are different for different spatial scales** of reference in which we find ourselves.*

*The theory so-called **Brane Cosmology** propose that the visible part of our four-dimensional universe is limited to a brane in a higher dimensional space called the **"bulk"** and also propose that the **Big Bang** emerged from a collision of branes. And it seems to explain the **weakness of gravity**, which could "leak" or escape to "bulk".*

3. SEARCHING FOR EVIDENCES

Any new scientific proposal must be adequately **proven (experimental) or demonstrated (theoretically)** to become officially accepted by the scientific community. And it is clear that to date there is no evidence (either experimental or theoretical) to prove or demonstrate this new model of the universe (**nD-Fractal Rainbow**).

The aim of this book is simply to propose an idea or approach to be considered, that provides a broader scenario that this we currently have about **our known universe**. To accept the feasibility of this proposal will facilitate to define and plan the appropriate experiments and formulations for prove or demonstrate it. On the contrary, if we just stick to the known, and to the currently accepted limits, it will difficult to leave them.

The most evident (experimental) way to prove this approach, would be to detect signals (waves) specific to these strips (inside and outside) of the spectral band of Our Universe.

For the **outer limits**, it seems more acceptable the possibility of detecting known external signals (waves), or some evidences of the existence of other entities or universes. And it seems possible that, someday, we may be able to detect certain **Electromagnetic Waves**, or the enigmatic **Gravitational Wave**s, income from outside of **Our Universe**. And also possible **Gravitational Forces** coming from other external or parallel universes (Brane Worlds).

If we assume that **Our Universe** is just one more of the millions of other possible universes, the potential **Electromagnetic and Gravitational Waves** generated by the **Big Bang** of these other universes (generated prior to ours), could be currently crossing Our Universe.

I disagree with those who say that these waves could never reach our universe, due to we're moving away from them faster than the speed of light. Because, it is also possible, that these waves may be generated previously, or simply we are not moving away so fast.

However, for the **inner borders** seems more complex its detection. Because the only <u>known</u> signals we could detect are the **Electromagnetic Waves** of wavelength less than the Planck size, and they have a very (too) high energy, or the <u>unknown</u> **Nuclear Waves**. Both possibly will have a very short scope, therefore their detection seem to be very difficult.

ELECTROMAGNETIC WAVES

A <u>known</u> signal that would allow us to validate the proposal are the **electromagnetic waves** that can interact in different spectral bands, from the smallest (<10 e-35 meters, the Planck length) to the largest (> 10 e +26 meters, the length of our universe).

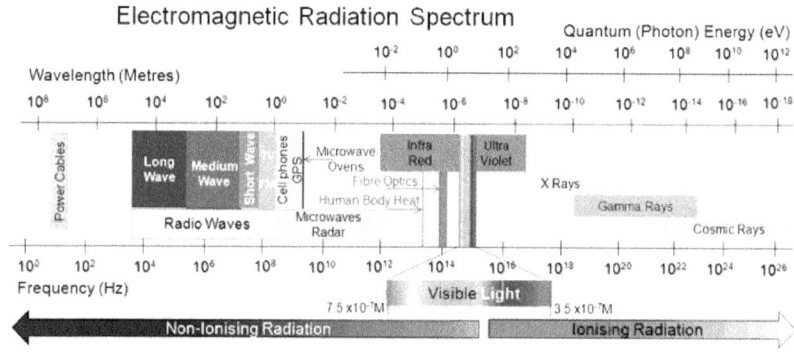

Fig.
6: Electromagnetic Radiation Spectrum
(10 e 0 to 10 e + 26 m and 10 e 0 to 10 e 26 Hz)

---

*According to String Theory EM waves cannot escape his own **brane**,and nor enter other branes, so this section will be meaning-less.*

*An EM wave from outside of Our Universe (Our Brane) would never enter it, so **neither of us could detect them.***

But the existence of these waves would this <u>mean that they were originated and come from outside the range of universe we know</u>? Any EM wave of wavelength less than 10 e -35 meters must necessarily come from scales smaller than this size ?, or could they also be generated within our known universe?. The same questions we could do on the EM waves above 10 e +26 meters.

If we imagine again the allegory that an **electron is the equivalent of the Earth** turning around the nucleus of an atom, like do the Earth around the Sun, and also that intelligent beings could live on this scale with technology equivalent to ours, would they be able to create FM radio waves? And if they were not, would they <u>be able to detect FM waves from higher scales</u>?

For EM fields of small dimensions, there is the problem of <u>high energy of these waves</u>, in which a single wave (photon) with the Planck wavelength (10 e -35 meters), carries energy of about 3 tons TNT. So to detect this type of electromagnetic radiation, we should hide in a shelter (bunker) and wait for it to produce a very loud explosion. Another problem is that these waves may have a <u>very short range</u> that will make it difficult its detection.

To detect large waves, also called **ELF ("Extreme Low Frequency")** or incorrectly "DC EM waves" ("Direct Current EM waves"), because they are nearly plane waves for our scale, we would have the problem opposite, their <u>low power</u>. In these waves the energy of the photons would approach zero, although its <u>scope could be very long</u>.

The electromagnetic waves of wavelength 10 e +10, that is a frequency of **30 Hz** (30 waves per second, nothing special in fact, similar to the magnetic resonance frequency of the magnetosphere field of the Earth), could be easily detected, measured and recorded with a **multi-meter connected to a strip chart recorder**, and with **very good insulation** to avoid any noise or interference. This may involve having to locate the meter and all installation in the interplanetary space (in a interplanetary satellite or spacecraft as far away from any object: planet, satellite,...).

But <u>things get complicated</u> if we want to detect and record **EM waves** of longer wavelengths, and especially when it comes to lengths **exceeding 10 e +25 meters**. This means a period of 1,000 million years for a complete cycle. And we should be able to perceive a small piece of the wave (accuracy 1/1.000.000.000 variation of the wave amplitude for 1 year of measurement and recording). And at least We would need several measurements (and so many years as measurements).

One of the difficulties encountered for detecting these low frequencies (**ELF: "Extreme Low Frequency"**) is the size of the antenna. The antennas should have a size approximate to half the wavelength with which they operate. However, there are other ways to build radio stations with smaller sizes due to **electrical lengthening**. But we would need a system of electrical lengthening between stations to very high distances (around the Earth, Earth-Moon, ...).

For long wavelengths of a few meters, a new generation radio telescope, the **Low Frequency Array (LOFAR)**, has started test operations in 2009 and **will be fully operational in 2013**. Most of the 40 planned stations operating in the Netherlands (http://www.lofar.org), six in Germany (http://www.lofar.de), and one in the UK (http://www.LOFAR-uk.org), one in Sweden (http://lofar-se.org) and one in France (http://www.lesia.obspm.fr/plasma/Lofar). The extension to other European countries are planned. Among the many possibilities LOFAR observation, is being able to track the radio synchrotron emission of low energy cosmic rays in weak magnetic fields. This allows us to observe the outermost regions of galaxies that are only accessible via radio waves.

But even if it were possible to detect these **EM waves of wavelengths so extreme**, positive or negative, is not 100% guaranteed that these waves came from beyond our known universe, but it is always a possibility to consider.

GRAVITATIONAL WAVES

Another option for detecting stimuli from outside of **Our Known Universe** could be the mysterious **Gravitational Waves**.

Gravity, which is the first force that the man knew (Galileo, Copernicus and Newton in the seventeenth century), before the **EM** (Oersted, Ampere and Maxwell in the nineteenth century) and obviously **Quantum forces** (weak and strong nuclear, Fermi and Yukawa in the twentieth century), nowadays is, possibly, the force (field) more unknown and enigmatic.

In physics, a **Gravitational Wave** is a ripple in space-time produced by a massive body accelerated. Gravitational waves are a consequence of general relativity theory of Einstein and transmitted at the speed of light.

On February 11, 2016, the LIGO Scientific Collaboration and Virgo Collaboration teams announced that they had made **the first observation of gravitational waves**, originating from a pair of merging black holes using the Advanced LIGO detectors.

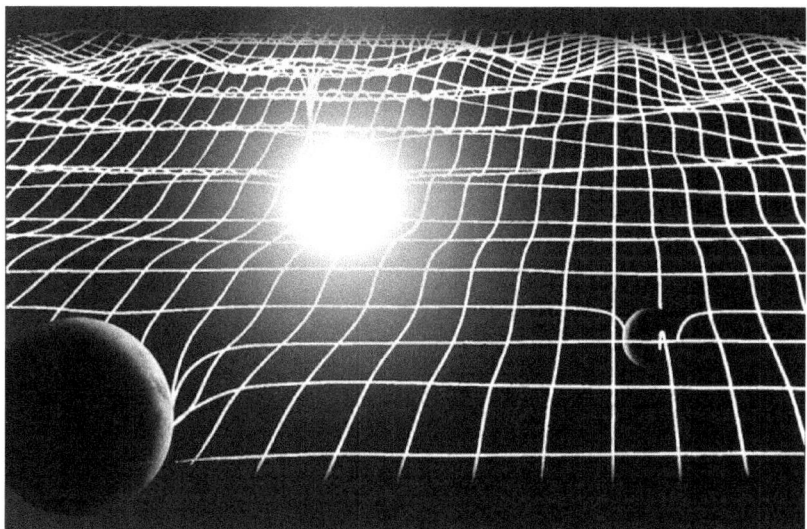

Fig.7: Gravitational waves

Gravitational waves can travel very long distances nearly un-changed, and its main possible sources are three types:

* **Catastrophic Sources by explosion**, produced by the coales-cence of compact binary star systems, or formation of neutron stars or black holes in supernovae.
* **Narrowband Sources:** Rotation of individual stars not asym-metric as binary stars far from coalescence.
* **Stochastic funds** due to the integrated effect of many sources of long distance, and even that could be generated by effects of the above sources produced in the early universe.

Latest studies suggest that one of the causes of **Dark Energy** might be the **Primordial Gravitational Waves** produced during the infla-tionary phase in the first moments of the Big Bang.

Currently there are several projects for the observation of gravitational waves, such as **LIGO** (USA), **TAMA 300** (Japan), **GEO 600** (Germany and UK), or **VIRGO** (France and Italy).

The pessimists believe that the actual detection of gravitational waves can only be done from space. A space mission called **LISA** ("Laser Interferometer Space Antenna") is currently under study to be the first **space observatory for gravitational waves** and could be operational around 2020.

The study of these waves will help us to answer questions about the early universe, the hypothetical end or their limits, and who knows if **other universes outside our**._It is possible that could be detected **Stochastic Backgrounds** generated by the Big Bang of other universes outside our own, or due to collisions between them.

NEW THEORY OF GRAVITY

*Different Fields of Forces of Nature we know (strong and weak nuclear, electromagnetic and gravitational) have associated a particle (Higgs, Gluon, Photon and Graviton, respectively) but the latter (the **Graviton**) is still to be demonstrated experimentally.*

*Currently it is still unknown how gravity really works (Newton, Einstein and Hawking, accepted that unknown how gravity works). Newton believed that gravity is an **attractive force** between masses), while Einstein proposed in his **Special Theory of Relativity** (the currently accepted officially) the concept of **curved space-time** as a possible cause.*

*Currently there are other proposals that try to explain gravity by **Quantum Theory**, such as **Arthur A. Larson** that propose it as an **interaction of emission and absorption of gravitons** from within the nuclei of atoms between them, producing self-movement toward each other atoms. Or, as **Hawking** said, "an exchange of gravitons between the particles constituting two bodies" that makes them move or attract. Gravity would be then quantified (based on particles) as they are all the other field forces of nature. A **quantized gravity is one of the holy grails of science**.*

*If the latter proposal (Larson and Hawking) was correct, would be a major change in the currently accepted scientific concepts about the universe, such as the **limitation of the speed of light**, the **Big Bang** and the **expanding universe**.*

EXTERNAL GRAVITATIONAL FORCES

According to Superstring Theory, the force of **Gravity is the only force that could escape from the world-branes** (because it is a closed string), and it could influence other near and parallel brane-worlds. Dark matter could be a manifestation of it. In contrast, the other three known forces (EM, S and W), can not escape our D-brane (being open strings).

Then, it could be possible that **Gravity forces coming from out-side of Our Universe** (from other "pocket" Universes or parallel universes) **could act over Our Universe and affect/influence over objects** inside (e.i. moving then on unexplained trajectories if we consider only the gravity forces of Our Universe).

MICROWAVES BACKGROUND RADIATION

Inflationary theory (if it was also valid outside the boundaries of our universe) predicts the **impossibility of traveling between two "bubble universes"** outside our own, because the space between them would be still in phase of inflation, expanding faster than the light can travel through it. So, any type of radiation that could prove the existence of this bubble, could not reach to us.

However, physicists have wondered what would happen if two of **those bubbles (universes) collide,** so that anywhere between them does not exist inflationary phase space, for example against our universe (our bubble).

A study and analysis of the Microwave Background Radiation of the Universe (made by NASA's WMAP) and treated with certain comput-er filters and algorithms, it seems that **detects collisions between other bubbles and ours** (Our Universe), in-primordial times. Such collisions would occur predictably **during the beginning of our universe**.

http://moriond.in2p3.fr/J12/transparencies/11_Sun-day_am/mcewen.pdf

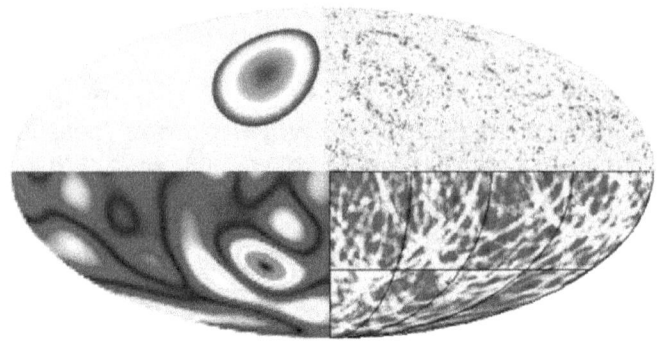

Fig.8: Possible collisions on the WMAP

This international team has created a new **computerized algorithm** to catch such irregularities in Our Universe. That they believe would be of a circular shape, and similar to temporary flattening that happens when a ball hits another.

Luckily, modern telescopes are able to study a faded snapshot of the universe when it was a baby: the **Cosmic Microwave Background**. The CMB is the radiation emitted by the hot plasma that dominated the universe to about 380,000 years after the "Big Bang", occurred more than 13 billion years ago.

But it could be that the current **maps of the CMB** (Cosmic Microwave Background) **are not precise enough** to detect small changes that presumably indicate a crash or collision inter-universal. So we're waiting for the new **Planck space telescope** data, which is recording the CMB with a **resolution three times better** than the most recent mapping CMB, created using the orbiter Wilkinson Microwave Anisotropy Probe (WMAP).

Data collection of Planck is scheduled to end 2012, but till now there are not definite conclusions about it.

UNIVERSES WITHIN THE PLANCK VOLUME

The **Size (Scale) or Length of Planck (ℓ P)** is that length below which it is expected that space will no have longer classical geometry. A lower measure of Planck Length can not be treated properly with current physics models due to the emergence of quantum gravity effects. To put it briefly, any <u>particle that is smaller such a</u>

length, will no longer have a classical geometry, ie, an object without the dimensions that we know (length, width, and depth).

> *The **Planck Length** is calculated as the distance traveled by a photon at the speed of light in the **Planck Time**, and depends on three fundamental constants, the speed of light (c), the Planck constant (h) and gravitational constant (G). The Planck length is about 10 e -35 meters.*
>
> *Furthermore, we can define different **Basic Planck Units** (length, time, mass, charge and temperature) and **Derived Planck Units** (energy, strength, power, density, pressure,...).*
>
> *The **Planck Volume** is the volume contained in a cube with edges of the Planck length (ℓ P cubed).*

Can there be something (and even more Universes) within this volume? For our brain it is very difficult to assimilate it. While the possibility of the existence of other universes outside of our own, though difficult but it is understandable, for lower dimensions, it is very difficult to understand and assimilate. Our brain makes us think that there must be an end, a minimum particle that until 100 years ago it was believed to be the **atom** (Greek "a (no)-tom (divisible)", which means **"no parts" or indivisible**), which was predicted by the former Greece but has not been demonstrated until the nineteenth century.

We must note that the size of Planck (10 e -35 meters) is 10 e -25 times smaller than the dimension of a standard atom (10 e -10 meters). Viewed in another way, for the Planck size is the size of an atom, what for us is the dimension of the Known Universe.

Very little is known of these dimensions, and concepts like **Quantum Foam** (basic fabric of the universe which is on these lengths), **or Quantic Fluctuations** (particles or radiations that apear from "nothing"), are confusing and fascinating concepts that are considered in these scales.

The **Theory of Kaluza-Klein** (1919-1926) proposed the existence of new spatial dimensions additional round that could be as small as the Planck length. And **String Theory** proposes the existence of six extra spatial dimensions to those known as **Calabi-Yau space**.

If confirmed the existence of these extra spatial dimensions beyond the Planck size, it would mean that, all those **entities or possible universes** that might exist **would be multi-dimensional**, and would be very different from our universe. And therefore, they would be very difficult to assimilate to our brain and from our reference scale.

What types of stimuli (waves) may come from these so small and multi-dimensional sizes? It is possible that at these scales below the Planck size (under 10 e-35 meters) there are **EM waves** (and possible also **waves of the strong and weak fields**), but these would be very small and short (both length in scope) and also very high energy. But it would also be possible that there may be **other types of force fields** currently unknown, and their waves, stimuli or signals, all undetectable for us and for the instruments currently available in our scale.

The technology system currently used by our scientists to study particles and phenomena such as small size, are the famous **particle colliders**. They try to observe and detect radiation and particles produced by colliding very small particles (hadrons, electrons, protons, ...) together and at very high speeds.

OTHER UNKNOWN FORCE FIELDS

As already mentioned, the **known Forces Fields are only four**: Weak, Strong, Electromagnetic and Gravitational. **Gravity** was the first force that the man knew (**XVII century**), before the **EM (XIX century**) and obviously the **Weak and Strong Nuclear** forces, (**XX century**). With these four force fields, are scientifically explained, currently, all observed phenomena in the known Universe.

What would happen if one or more **new force fields were discovered**? Is this a viable hypothesis?

Obviously, this is a very feasible possibility. If not it would be such as to say that we know almost everything in physics, and that few things we have left to discover. Suppose this would be very pretentious and even irresponsible for serious and rigorous scientific. Especially considering that, four centuries, ago there was not any known. Two centuries ago was known EM, and only 100 years ago were known the weak and strong.

As we have seen, these known forces are more prevalent in some scales than others. Then maybe these other force fields to know presented with greater splendor beyond the known limits of our universe (either in the positive or in the negative). But, if we do not know them, more difficult will be to detect, but not impossible. We simply have to keep an open mind to this possibility and to be alert to any signs or abnormal/strange phenomenon that can not be explained with the known fields.

TABLE 1: COMPARATIVE MODELS (MAINSTREAM & nD-FRAC-TAL RAINBOW)

MAINSTREAM MODEL	nD-FRACTAL RAINBOW MODEL
The **Universe** (is unique) is limited to the upper scales by the space generated from the **Big-Bang** (10^{+27} meters), and to the lower levels by the **Planck** scale (10^{-35} meters) in which the appearance of **Quantum Gravity** prevents that a lower measure can not be treated properly with current physical models.	**Our known Universe** (between 10^{-35} & 10^{+27} meters) is not more than a **small band within the total Global Universe scale spectrum**, which, in principle, can be infinite. And as physical science and technology advance, it will widen its stretch.
The **Universe** (which is unique) began in a **Big Bang** 13,700 million years ago and it is expanding **isotropically** since that time.	**Our Universe** is a **"bubble"** else of the many that exist in a larger scale than ours within the **Global Universe**, and which also had (or will have) their own **Big Bang** at different times.
In the **Universe** there are four **Forces Fields** (strong, weak, electromagnetic and gravity) which explain all known physical phenomena.	**Out of** the current limits of this range scale of **Our Universe** may be **other Force Fields** acting very weakly in Our Universe.
Underlying Fundamental Laws of the **Universe** are the same for the whole universe even that they can manifest differently in the small (Quantum Model) and the large (Relativistic Model).	While we accept that the **Underlying Fundamental Laws** are the same for all spectra and ranges of the **Global Universe**, they manifest differently through them, requiring different models that formalize and explain this different zones.
The **Universe**, currently, can be explained and standardized by two physical models: **Quantum Model** (for small) and **Relativistic Model** (for large). And is evolving into a pattern that would unify all previous models (**M-theory or Super-string**), covering the entire universe.	For the different bands of the **nD-Fractal Rainbow** of the **Global Universe**, may dominate different models and patterns to better explain its phenomena (**Relativistic and Quantum Models**). The **M-theory** may simply cover a wider range of zones, encompassing the previous models.
The **Universe** is unique, and all waves and stimuli, and the fields and forces we know, are what conform the whole. Only would be possible other **Parallel Universes**, and potential living beings, in other **Spatial Dimensions.**	For the different bands of the spectrum of the **3D Rainbow** may dominate different stimuli and waves, different fields and forces. And can be formed different entities and bodies. **Parallel Universes** can coexist, and even different living beings in different **Scale Levels**.
Our species is <u>halfway</u> between the largest structures and the smallest of the **Universe**, and we are very lucky because it serves as a platform to observe the universe in all its aspects.	Our species is <u>halfway</u> between the largest structures and the smallest of the **known universe**, because from the reference spectrum scale where we are, <u>that is what our technology and expertise allows us</u> to observe and learn towards the two directions.
The **Universe** is considered a closed physical system in which **the energy contained in it is constant**.	Different **levels of scales** can be considered **Physical Open Systems**, and this would mean that there could be an **exchange of energy between them**.

47

TABLE 2: POSSIBLE SIGNALS BEYOND THE LIMITS OF OUR UNIVERS

POSITIVE SCALES

Electromagnetic waves of very high wavelengths (above 10 e +30 meters) that cross Our Universe.

Gravitational Waves from Stochastic Funds generated by the Big Bang of other universes outside Our Universe.

Gravitational Forces coming from other external or parallel universes (Brane Worlds).

Microwave Background Radiation of the Universe treated with certain filters and computer algorithms to detect **collisions of other bubbles** out of (or against) Our Universe.

Other waves generated by **unknown interaction fields** (from other universes), that prevail only on scales very high (above 10 e +30 meters), but their influence can be very weak in our scales.

NEGATIVE SCALES

Electromagnetic waves of very low wavelengths (bellow 10 e -35 meters), although its scope can be very short and difficult to detect.

Potential Waves generated by **Weak and Strong Fields**.

Other waves generated by **unknown interaction fields**, that prevail only on scales very low (bellow 10 e -35 meters), but their influence can be very weak at higher levels.

PART II

This second part of the book include the second and reviewed article: *"The Scale Landscapes (Relativity) of the Universe"* (David Piñana, October 2015).

The objective of this second article was:

- To **contrast the proposals of the first article (Part I) with the latest advances in cosmology** by other prestigious physicists and cosmologists.
- To **give a broad overview of the latest different cosmological theories**, as well as the latest developments and also the future lines of study.
- I noted that **these "leading" theories largely supported the scale proposal (ND-Fractal Rainbow) of my first article**. Although these were from partials & different points of view (emergency, scale relativity, cosmic landscapes,…).
- Furthermore, this second article proposes a **different view of certain concepts, theories and physical principles** (Dark Matter and Energy, Uncertainty Principle, …), which can be explained in another way if we rely on the scale proposal of the Universe.

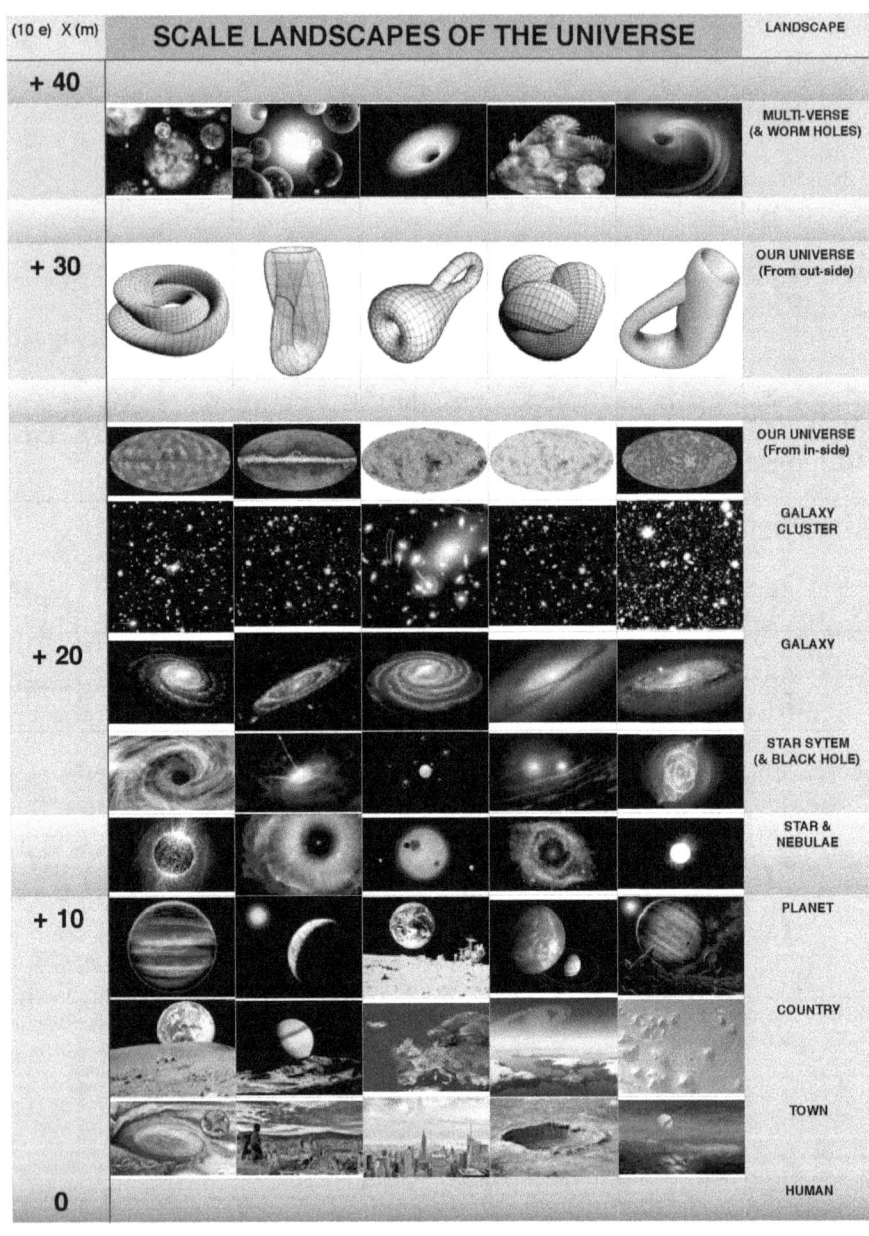

Fig.9: Positive Scale Landscapes

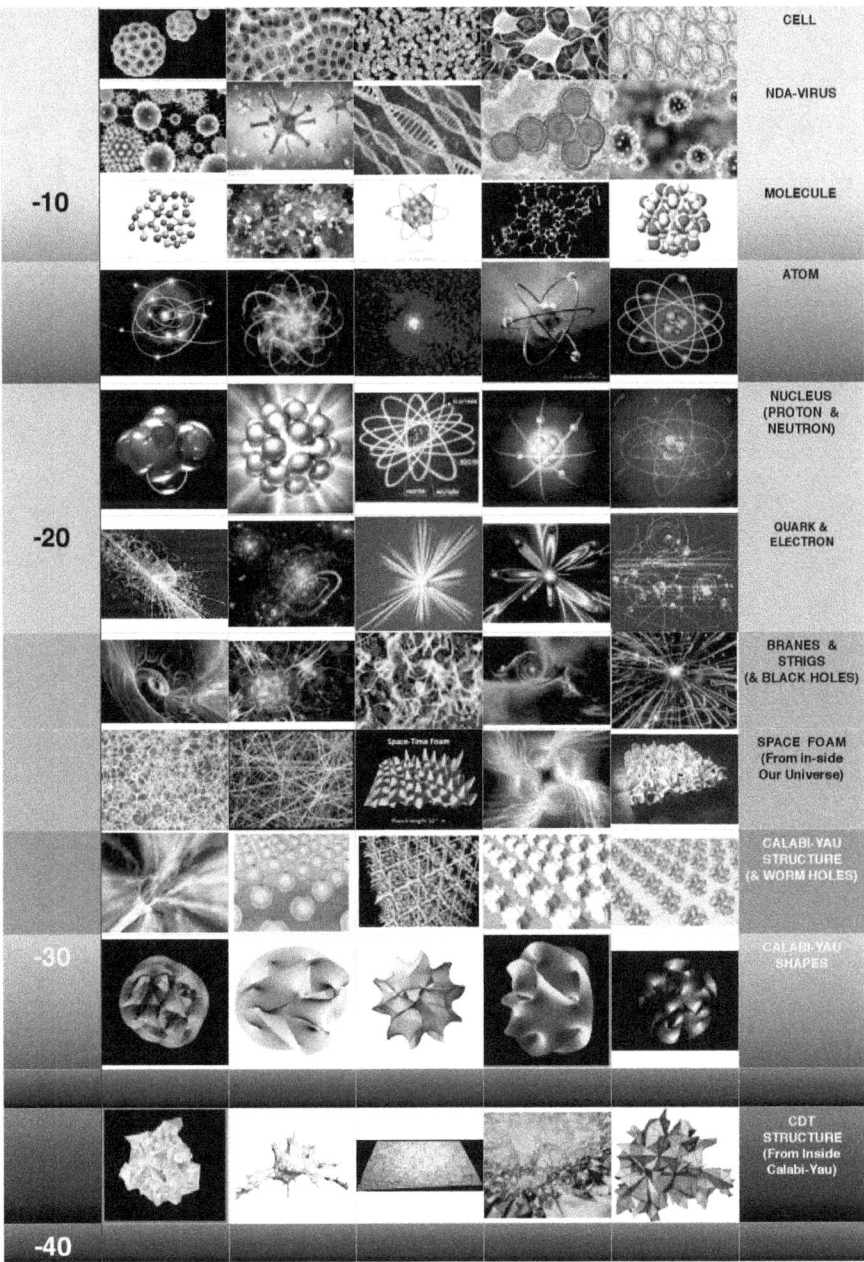

Fig.10: Negative Scale Landscapes

4. UNIVERSE LANDSCAPES

The term LANDSCAPE was established by L. Susskind to describe the upper spectrum of our universe: Cosmic Landscape. But it will be also used in this article to name the different scale/size levels (landscapes) of the universe **(see Fig.11).**

So we can define different LANDSCAPES that describe different scale/size levels with their own concepts and laws:

- **Newtonian Landscape**
- **GR Landscape**
- **QM Landscape**
- **Etc.**

And we could also forecast other LANDSCAPES that describe further scale/size spectra:

- **Cosmic Landscape**
- **Planck Landscape**
- **Supra-Cosmic Landscape**
- **Infra-Planck Landscape**
- **Etc.**

In this book **the Supra-relativistic, Cosmic and Planck Landscapes will be discussed in more detail** in the following chapters, whereas now we will only make some comments on the rest of landscapes, due to they are better known and there is already a lot of literature dealing with them **(Landscapes: Newtonian, Relativistic, Quantum, Chemical, ...)**. The first three are summarized in **ANNEX 6 (Mechanical Theories).**

The human (and therefore his brain as his senses) has evolved into the **Newtonian Landscape**, and, as much, we are able to observe and understand the events that occur within the range between 10 e -10 m to 10 e +10 m.

In this scale spectrum (landscape) is where the theories and **laws of Newton and Maxwell** work perfectly, and even without knowing them, humans have been able to evolve over thousands of years.

Within this spectrum, and **included in the laws of Newton and Maxwel**l, other subjects are developed that are developed mainly in the **Faculties of Engineering**: Fluid Mechanics, Structures, Electrotechnics, Telecommunications, Thermodynamics, etc.

And towards the smaller scales we find the **Biology** that deals with living animals: cells, viruses, DNA, ...

But during the last (100-200) years, we have been expanding our knowledge range for both: positive scales 10 e +20 m (**Relativistic Landscape**) and negative scales 10 e -20 m (**Chemistry Landscape**).

- The **Chemical Landscape** presents the basis of the composition of matter in atoms and molecules, and their transformations and combinations. And on the smaller scales (10 < e-15 m) the very formation of atoms from their components (protons, neutrons, electrons, ...).
- The **Relativistic Landscape** (typical of Einstein's Theory of Relativity) is required for large scales (> 10 e +10 m) in which distances and masses are very large (and given the limitation of the speed of light), there make it necessary to consider the Relativity of Time.

We are currently expanding these spectrums at major and minor scales:

- > 10 e+20 m to approx. 10 e +25 m (**Supra-Relativistic Landscape**) where MOND Theory and/or Dark Matter & Energy concepts apear, but we still do not fully understand.
- < 10 e-20 m to approx. 10 e -30 m (**Quantum Landscape**) where we have QM, QED, QCD and QG Theories, and also Uncertainty Principe and particle wave function concepts, that we also still do not fully understand.

In the **Supra-relativistic Landscape** (above 10 e+20 m) we observe that the Galaxies and the Clusters of Galaxies do not follow the behaviors that would be expected if they followed the Laws of Newton or Relativistas, reason why they appear the concepts of Matter Dark or alternate laws of MOND, MOG, TeVeS, etc., which we will see in the following chapters.

Fig.11: The Landscapes of the Universe

10 exp X	MODELS (PHYSIC / MATHS)	SCALAR LANDSCAPES	FORCE FIELDS						E-DIM	T-DIM	CONCEPTS
			G	EM	S	W	X	Y			
50											
		SUPRA-COSMIC LANDSCAPE							4-9D	1/0	
40											
											PRIMIGENIUM FIELD
30	SUSSKIND	COSMIC LANDSCAPE							4-9D	1/0	INFLATON FIELD
											HIGGS FIELD
		SUPRA-RELATIVISTIC LANDSCAPE									DARK ENERGY
20											DARK MATTER
	RIEDMANN										GALAXIES
	EINSTEIN	RELATIVISTIC LANDSCAPE									BLACK HOLES I
10	FRIEDMANN										SOLAR SYSTEM
											SUN
											HEARTH
	NEWTON										TOWN
	EUCLIDES	NEWTONIAN LANDSCAPE							3D	1T	HUMAN
	MAXWELL										CELL
											NDA
-10											MOLECULE
											ATOM
	BOHR	CHEMISTRY LANDSCAPE									PROTON
-20	DIRAC-NEWMANN										QUARK-ELECTRON
	FEYMANN										BLACK HOLES II
	HEISENBERG	QUANTUM LANDSCAPE									UNCERTAINTY PRINCIPLE
-30	SCHRÖDINGER										QUANTUM FLUCTUATIONS
-35	PLANCK	PLANCKIAN LANDSCAPE							6D		CALABI-YAU 6D
-40											
		INFRA-PLANCKIAN LANDSCAPE							3-5D		
-50											
-60											
	CARLIP	CDT LANDSCAPE							2D		

Vertical model/theory labels (left columns): RG, CM, BQ, Q, QED, QCD-QG · TERMODINÁMICA · M TEORY · HOLOGRAM · FRACTAL · CAUSAL DYNAMIC TRIANGULATION

GR GENERAL RELATIVITY
MC CLASSIC MECHANIC
BQ BIO-CHEMISTRY
Q   CHEMISTRY
QED QUANTUM ELECTRO DYNAMIC
QCD QUANTUM CHROMO DYNAMIC
QG QUANTUM GRAVITY
GR GENERAL RELATIVITY

On the other hand, the **Quantum Landscape** presents the landscape that appears within the smallest elementary particles (proton, electron, neutron, neutrino, ...) and its components (Quark, Boson, ...), as well as the forces that (Nuclear Strong and Weak).

Beyond these landscapes, we may forecast other new scale landscapes that we are just beginning to discover, and where their new laws and concepts we are just trying to understand: **Cosmic Landscape** and **Plank Landscape.** In the following chapters we will talk about them.

5. COSMIC LANDSCAPE

In his book **The Cosmic Landscape (2006) Leonard Susskind** presents the Cosmic Landscape concept as a place (landscape) beyond Our (known) Universe to a higher level or dimensional scale (from 10 e +27 to 10 e +40 m).

The Cosmic Landscape would be a scale spectrum above Our Universe (**"Bulk"**), where, possibly, and according to string theory, would have superior spatial dimension (4D-9D-25D ?), and where would "float" (coexist) other universes (**"pocket universes" or World-branes**) with other spatial dimensions (N-dimensional), in various stages of development (expansion, implosion,...) and with different physical constants (Cosmological, G, h, c, ...). And where **Our Universe would be only one among many other possible universes** (up to +10 e +500 according to mathematical possibilities offered by the string theory).

This cosmic landscape (with X-space dimensions) would contain a _**Primordial Field"**_ (energetic field) that would have different values at each point of the X-dimensional space (forming different energy valleys, with different "space-phase"), It gives us a view of N-Dimensional valleys and mountains which give the name of Landscape. One of these valleys would be Our Universe, where the value of Primordial field would be our **Cosmological Constant (Λ=10 e -116 J).**

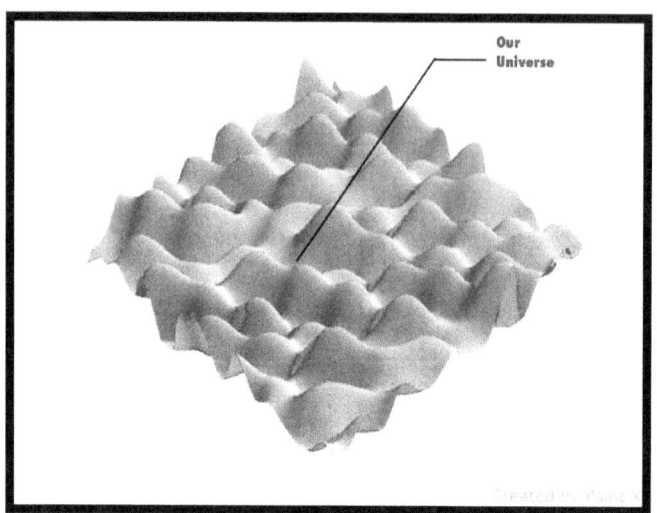

Fig.12: Cosmic Landscape (3D representation)

The Time (we will talk later on in detail about Time concept) could be a dimension that depends on the entropy (**S**), being positive if $\triangle$**S** is positive, and possibly negative, if $\triangle$**S** is negative (which would contradict the Second Law of Thermodynamics). **We cannot ensure that there is any time in the Cosmic Landscape, and if it would be positive, negative or null**. Possibly it would also depend on the $\triangle$**S**, or other new emergent concepts.

According to S.Hawking, the **Big Bang** of Our Universe **emerged from nothing** through the quantum vacuum fluctuations, starting from the spontaneous creation of particles and fields (matter-antimatter and matter-gravity, and also from other particles and fields). This particles and fields prospered through inflation, and retaining the initial energy (= zero ?), by considering the matter as positive energy, and gravity (or the gravitational energy) as negative energy, counteracting both, and making the total universe energy zero (as before the Big-bang was). And it could also have happened with the other particles and forces (being positives-negatives energies).

Under the **Cosmic Landscape proposal,** it would be an alternative that the **Big Bang** occurred from this primordial energy field, previously existing in the Cosmic Landscape. And the energy of Our Universe could come from this preliminary energy. Big Bang could happen as a decrease of this primordial energy, creating an energy valley and creating the conditions for a short burst of inflation, transforming this primordial energy (**inflaton energy ?)** field in matter and radiation (and transforming the primordial space (9D ?) to our 3D space (more other 6D with very small size: KK). From there we can follow the same reasoning of Hawking proposal, although the **initial energy would not necessarily be zero as Hawking propose.**

The Big Bang could be due to an inflationary (energy and dimensional) explosion that would generate, from the Bulk Primordial Field (nD dimensional), the 3D space of Our Universe with an intrinsic energy (matter).

Some scientists have proposed that Our Universe (3D space) could be like a "White Hole" inside a 4D universe ("bulk").

"From a cosmological point of view, the entropy of Our Universe is always positive. It means that long time ago (in the Big Bang) Our Universe should be very low entropy (near to a state of perfect order).
How something so orderly, could lead to the expansion that formed Our Universe, making it increasingly chaotic ?"

("Before the Big-bang", Martin Bojowald, 2009)

ENERGY-MATTER OF OUR UNIVERSE

According to the **Principle of Conservation of Energy (or First Law of Thermodynamics),** the total amount of energy of any isolated or closed physical system (without interaction with any other external system) remains unchanged over time. But that energy can be transformed into another form of energy. In short, the Law of Conservation of Energy states that energy can neither be created nor destroyed, it can only change from one form to another.

If this principle was always valid (for any moment of the life of the universe) and for the entire universe created after the Big Bang, it would imply that the **total energy of the (our) universe would be the same now than it was in the beginning, and the same that it will be at the end.**

Where does it come the total energy of the (our) universe?

According to Stephen Hawking, the Big-bang (and the whole Energy and Matter of Our Universe) began (from Nothing: zero energy) due to quantum vacuum fluctuations by creating positive energy and negative energy (ie. Matter and anti-matter). **So now (and always) the total energy of the universe is zero,** and, for example, mass and gravitational energy cancel each other (as well as the other forces and particles of the universe could cancel one to each other).

Another proposal/option could be that **energy is intrinsic to the space**, and that **energy is created (increase) when space is created (enlarged or expanded)**. But this proposal breaches (violates) the Principle of Conservation of Energy (or First Law of Thermodynamics). It will mean that the **Energy of Our Universe increase at the same time that the space volume of Our Universe is enlarging (expanding).**

For these both proposals/options, we can state that: *"The Known Energy-Matter (and space) that we see and detect in Our Universe might be considered as an emergent concept arose after the Big-bang".*

On the other hand, considering the **Cosmic Landscape** proposal, Big Bang of Our ("pocket") Universe would only be an event of the many that occurs within the whole universe (Cosmic Landscape), and the energy (matter) of Our Universe could come from a prior energy already existing on the Cosmic Landscape (**Primordial Field**), where different "pocket" universes would be created and destroyed as bubbles in a glass of soda gas. The dilemma for the energy of Our universe disappears, but it raises to a higher level: **Where does it come Cosmic Landscape (Whole Universe) energy-matter ?.** But we could try to solve this second problem afterwards, when we understand better Our Universe and the Cosmic Landscape.

It would be like the prisoners inside a cave who could only know what happens outside through the shadows produced by the light of an outside fire that reflected through a small hole in the wall. His vision would be only that of 2D shadows on the wall.

If they could ever leave the cave, the perception and understanding of the outside would greatly improve, and if they could also see the Sun, their perception would be very different, even if they still did not understand everything.

We have been describing different proposals about from where it came or how it was created the Energy-Matter (**Known Energy-Matter**) of Our Universe, but recent observations tell us that all the **Energy-Matter we know is only approx. 5 % of the Whole Energy-Matter of Our Universe**. The other 95 % are Dark Energy (70 %) and Dark Matter (25 %).

This can give us an idea of how little we really know about the universe, and how far we are from a (physical-mathematical) scientific model that describes and parameterizes it properly.

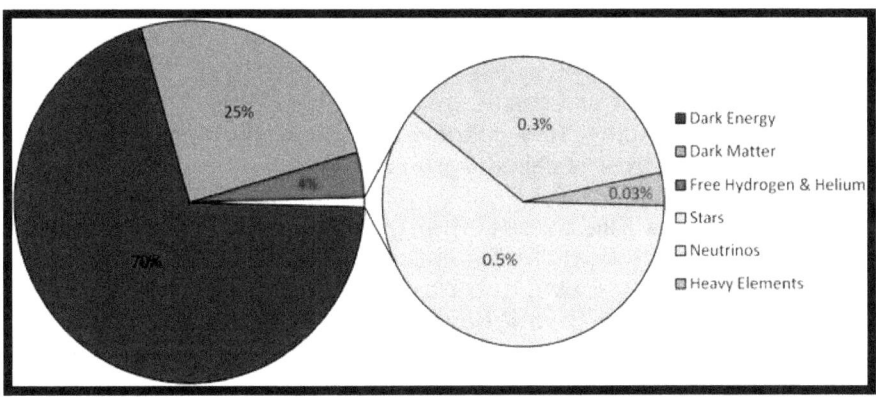

Fig.13: Matter-energy contents of Our Universe

This can give us an idea of how little we really know about the Universe, and how far we are from a scientific (physic-mathematical) model that correctly describes and parameterize it. It is often said that we know only 5% of the Universe, but according to the proposal RAINBOW FRACTAL should say that ***the universe becomes unknown as we move away from our scale, both to large and to the small.***

We can summarize the following hypothesis about the energy of Our Universe:

1.-Considering **Our Universe as a CLOSED SYSTEM**:

1.1.-The **total amount of energy in Our universe is exactly zero**: its amount of positive energy (e.i: matter) is exactly canceled out by its negative energy (e.i: gravity).

1.2.-If dark energy is an intrinsic property of empty space (as hypothesis) then **energy is created when space expands.**

2.-Considering **Our Universe as an OPEN SYSTEM:**

- The **total amount of Energy of Our Universe is the result of the Energy that goes into it** (Big-Bang, Gravity coming from outside,...?), **and the energy that is leaving it** (Gravity leaving outside, Black & Warm holes,...?)

3.-Any way the UNIVERSE is considered as an OPEN or as a CLOSED SYSTEM: **Energy concept is not well-defined in GR Theory.**

DARK ENERGY

Dark Energy tries to explain the **acceleration of the (isotropic) expansion of Our Universe.** If, as it seems, the Universe is expanding isotropic and accelerated, it implies that something is transmitting an energy for it is thus, if it would not it will be static and without expanding. It is unknown what is and where this energy comes from, and by it is called Dark Energy.

Dark Energy **acts like a repulsive gravity.** It is (possibly) caused (Option 1) by some kind of different matter we know (anti-matter?). Known Matter is attractive, and **Dark Matter is repulsive: exerts a negative pressure throughout the space**.

But the simplest explanation (Option 2) for Dark Energy could be that it is simply the "cost of having space": that is, **a volume of space has some intrinsic fundamental energy**. As we will see later, empty space (vacuum) is empty of matter but not completely empty (vacuum is not the same than nothing).

Vacuum energy is an **underlying background energy that exists in space throughout the entire Universe** (nD). Vacuum energy is expected to contribute to the cosmological constant, which affects the expansion of Our Universe.

The effects of **vacuum energy can** be experimentally observed in various phenomena such as spontaneous emission, the Casimir effect and the Lamb shift, and are thought to **influence the behavior of the Universe on cosmological scales.**

DARK ENERGY = VACUUM ENERGY... (?)

Dark Energy value is the **Cosmological Constant** (nowadays it is considered as independent of time and place in our universe, Λ = **10 e-116 J**).

DARK ENERGY = COSMOLOGICAL CONSTANT

Other possibility (Option 3) could be that **we might consider that the accelerated expansion of Our Universe itself is produced by the spontaneous generation of new space** (newly created, or coming from outside of our universe: Cosmic Landscape or "Bulk"). This option would be in line with the Big-Bang of Our Universe theory as **one more of all Big-Bangs that occur in the Cosmic Landscape.**

And as stated in the previous chapter, **the expansion of space, could also lead to an increase in total energy of Our Universe.**

---

*But it would also be possible that the value of the **"cosmological constant" has varied (increased-decreased) from the Big-bang**, depending on the rate of speed (acceleration) of the expanding universe.*

DARK MATTER (& OTHER THEORIES)

Dark Matter is just a physical concept that most mainstream scientists propose to explain rotation speed of galaxies that do not follow Newtonian and Relativistic models. Dark Matter proposal forecasts other matter that we cannot see and we only can detect by this galaxy phenomenon. Dark Matter represents four times more matter than the known matter. To date there are many theories about Dark Matter, but no one could prove or show their existence (WIMPs, Axions,...). To find these new particles is currently one of the most important subjects of study and research (HLC-CERN, ADMX, DAMA,...).

> *According to Newton and Einstein's Theories, bodies that revolve around one another, like planets around the Sun, spin slower as they move away from each other. Venus rotates faster around the Sun than Earth, but Uranus slows it down to Earth.*
> *But this is not the case with galaxies where all stars (more or less away from the center of rotation) rotate at the same speed (as if they were dots on a rotating plate).*
> *In order to understand this phenomenon it was proposed that there should be more energy than we see among the stars of the galaxies.*

But another way (<u>Alternative 1</u>) to solve this problem (the galaxy rotation) is to assume that **Newton and GR theories, are not valid for very long distance scales (> 10 e +20 m).**

MOND Theory and their last version TeVeS gives other alternatives to the same problem, whereas the laws of Newton and Einstein (SR & GR) vary with the distance/size scale. <u>See Annex 1.</u>

*"In physics, **Modified Newtonian Dynamics (MOND)** is a theory that proposes a modification of Newton's laws to account for observed properties of galaxies. **<u>Created in 1983 by Israeli physicist Mordehai Milgrom,</u>** the theory's original motivation was to explain the fact that the velocities of stars in galaxies were observed to be larger than expected based on Newtonian mechanics. Milgrom noted that **this discrepancy could be resolved** if the gravitational force experienced by a star in the outer regions of a galaxy was proportional to the square of its centripetal acceleration (as opposed to the centripetal acceleration itself, as in Newton's Second Law), or alternatively **if gravitational force***

came to vary inversely with radius (as opposed to the inverse square of the radius, as in Newton's Law of Gravity). ***In MOND, violation of Newton's Laws occurs at extremely small accelerations,*** *characteristic of galaxies yet far below anything typically encountered in the Solar System or on Earth."*

"And Tensor–vector–scalar gravity (TeVeS), developed by **Jacob Bekenstein in 2004,** *is a relativistic generalization of Mordehai Milgrom's Modified Newtonian dynamics (MOND) paradigm."* And this theory best explains what happens at scales higher than the Galaxies (Galaxies Clusters), where theories of Newton, Einstein and also of MOND do not allow to explain.

We should see if the laws of Newton-Einstein suffer variations (changes) depending on the spatial scale of reference **(See Annex 1: MOND Theory) :**

- **Newton's (& Einstein) law** *seems to work perfectly to explain the dynamic phenomena at* **scales up to the Solar System (max. 10 e +15 m),,.**

- *From this distance we must consider the* **MOND´s law,** *which should fully explain the dynamic phenomena at scales between that* **10 e+15 to 10 e +20 m (galaxies) ,** *specifically, the rotation of the stars from a distance to the center of the galaxy.*

- *Should be* **above these scale (10 +20 m),** *to the* **ends of Our Universe (10 +27 m),** *where the* **other dynamic laws (TeVeS´ law)** *could describe better these scales phenomena.*

- *And* **above these scales (> 10 e +30 m)** *other* **new emergent laws** *will apear to explain new emergent phenomena that we cannot forecast by now.*

Other proposal (Alternative 2) could be the **Fractality of Spacetime**:

"Applying relativity to fractal non-differentiable spacetime, Laurent Nottale, in his **Scale Relativity theory,** *suggests that potential energy arises due to the fractality of space, and accounts for the missing mass-energy observed at cosmological scales".*

This missing of mass-energy could make also that **G (gravity constant) decrease at cosmological scales**. Gravity could be missed through the fractal space-time. *(see Annexes 1 & 5).*

The **Dynamic Laws of Physics (and Universal Gravitation)** *have varied over time, and even Einstein had already proposed that they still has to evolve:*

ARISTOTLE: $\qquad\qquad$ *F = m.v*
NEWTON: $\qquad\qquad$ *F = m.a*
EINSTEIN. $\qquad\qquad$ ***E = m.c^2 (*)***
MOND: $\qquad\qquad$ *F = m.a.(A/A$_0$)*
FRACTAL RAINBOW: $\quad$ ***F = f (scale) = m.a.(scale factor)***

Or better G (Gravity Constant) vary with the scale/distance due to fractal space-time: G = f (Scale/distance factor)

(*) This equation does not correspond to the same dynamic concept but has many similarities.

Another possible (<u>Alternative 3</u>) **explanation to the Dark Matter** (of which I do not know if there is any current theory at the moment that considers it), could be that **for these large scales emerge other forces or interactions (Y) due to the accumulation of stars and/or galaxies** that generate effects orbehaviors unanticipated. In the same way that the Force of Gravity arises when many atoms and molecules group together forming large massive bodies (asteroids, planets, ...).

Another alternative for Dark Matter could be that for larger scales emerge new forces (interactions) unknown nowadays.

6. PLANCK LANDSCAPE

According to the Uncertainty Principle, the value of a field and its temporal change rate (waves) plays the same role as the position and velocity of a particle. And an important consequence is that **there is "NO" a vacuum**. Since the empty space means that the value of a field is exactly zero and the rate of change of the field is zero (if it were not so, space would not be empty). Then as the Uncertainty Principle does not allow exact values have both simultaneously, the space is never empty. You can have a **minimum energy state, called "vacuum energy"** and is subject to what is called *vacuum quantum fluctuations*, consisting of particles and fields that appear and disappear from existence. As virtual particles, they cannot be observed by particle detectors, but they can be by indirect methods (energy changes in electron orbits).

> *The Planck length is related to Planck energy by the uncertainty principle.*

The nature of reality at the Planck scale is the subject of much debate in the world of physics, as it relates to a surprisingly broad range of topics. It may, in fact, be a fundamental aspect of the universe. In terms of size, the Planck scale is extremely small (many orders of magnitude smaller than a proton). In terms of energy, it is extremely 'hot' and energetic. The wavelength of a photon (and therefore its size) decreases as its frequency or energy increases. The fundamental limit for a photon's energy is the Planck energy. This makes the **Planck scale a fascinating realm for speculation by theoretical physicists** from various schools of thought. Is the Planck scale domain a seething mass of virtual black holes? Is it a fabric of unimaginably fine loops or a spin foam network? Maybe at this fundamental level all that remains of space-time is the causal order? Is it interpenetrated by **innumerable Calabi–Yau manifolds, which connect our 3-dimensional universe with a higher-dimensional space?** Perhaps our 3-D universe is 'sitting' on a 'brane' which separates it from a 2, 5, or 10-dimensional universe and this accounts for the apparent 'weakness' of gravity in ours. These approaches, among several others, are being considered to gain insight into Planck scale dynamics. **This would allow physicists to create a unified description of all the fundamental forces.**

> *"The incompatibility between **GR and QM can only be solved** if we reject the notion that space is a fundamental concept, and **we accept that space emerges from the expansion of the Universe itself, and space is an emerging concept**"* (Lee Smolin).

CASUAL DYNAMICAL TRIANGULATION

Near the Planck scale, the structure of space-time itself is supposed to be constantly changing due to quantum fluctuations. CDT theory uses a triangulation process which varies dynamically and follows deterministic rules, to map out how this can evolve into dimensional spaces similar to that of our universe.

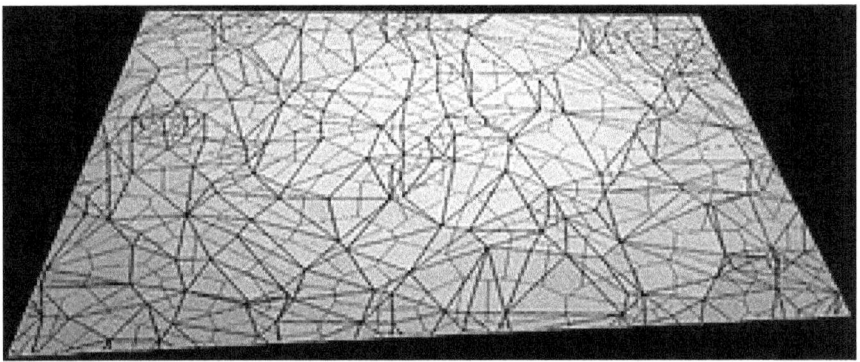

Fig.14: CDT Planck Landscape texture

Causal dynamical triangulation (abbreviated as CDT) invented by Renate Loll, Jan Ambjørn and Jerzy Jurkiewicz, and popularized by Fotini Markopoulou and Lee Smolin, is an approach to quantum gravity that like loop quantum gravity is background independent. This means that it does not assume any pre-existing arena (dimensional space), but rather attempts to show how the spacetime fabric itself evolves. At large scales, it re-creates the familiar 4-dimensional spacetime, but it **shows spacetime to be 2-**

d near the Planck scale, and reveals a fractal structure on slices of constant time. **These interesting results agree with** the findings of Lauscher and Reuter, who use an approach called **Quantum Einstein Gravity, and with other recent theoretical work.**

*"It is widely accepted that, at the very smallest scales, **space is not static but is instead dynamically-varying**. Near the Planck scale, the structure of spacetime itself is constantly changing, due to **quantum fluctuations**. This theory **uses a triangulation process which is dynamically-varying** and **follows deterministic rules, or is dynamical, to map out how this can evolve into dimensional spaces** similar to that of our universe. The results of researchers suggest **that this is a good way to model the early universe, and describe its evolution.** Using a structure called a simplex, it divides spacetime into tiny triangular sections. A simplex is the generalized form of a triangle, in various dimensions. A 3-simplex is usually called a tetrahedron, and the 4-simplex, which is the basic building block in this theory, is also known as the pentatope, or pentachoron. Each simplex is geometrically flat, but simplices can be "glued" together in a variety of ways to create curved spacetimes. Where previous attempts at triangulation of quantum spaces have produced jumbled universes with far too many dimensions, or minimal universes with too few, CDT avoids this problem by allowing only those configurations where cause precedes any event. In other words, the timelines of all joined edges of simplices must agree."*

From: http://everything.explained.today/ Causal_dynamical_triangulation/

SPACE AND VACUUM

Conventionally we mean by empty space (vacuum) that space (volume) where there is no matter (atoms, molecules, ...).

Both from the Newtonian point of view (space is absolute with fixed coordinates) and from the Relativistic one (space is curved), space is filled with matter (in the form of gas, liquid or solid) and/or energy (waves , ...). When we take out all the matter, there is a vacuum. But **emptiness is not nothingness.**

It is not the same "vacuum" that "nothing"

The **"classical" vacuum as such does not exist.** There is talk of the **"quantum" vacuum**, which is not really empty but full of particles, antiparticles and photons (energy) in constant oscillation. Photons create particle-antiparticle pairs, which annihilate each other by generating photons, and thus in a cycle without beginning or end (**"quantum foam"**).

As we can read in the summary of Frank Wilczek's article, the **vacuum is formed of "something"**, the vacuum is like a substance formed by extremely small elements that we do not know. That is, the space between an electron and a proton in an atom, an empty space because there is no matter, it actually has life by itself and it is formed of "something", it is not true that there is nothing. And as we have seen has energy: **vacuum energy.**

In the article *"What´space" by Frank Wilczek (Mit physics annual 2009)*, we can read:

*"Space is effervescent, substantial, weighty, and elastic. Each of these properties equates to specific, observable phenomena; they are not whimsical metaphors. **Space has a life of its own, and exists independent of any matter that might occupy it.** Indeed, in our most fundamental equations **particles**—the building blocks of matter—**are described as disturbances in the activity of** space-filling fields, or in other words of **space itself**.*

*Modern quantum physics brings in ideas of a different order. **Quantum reality lives in spaces whose meaning, size, and structure transcends classical ideas about physical space.** To get in tune with Nature, we must vastly expand our conceptual universe.*

***The structure of space is encoded in the metric field.** Like all fields the metric field is subject the laws of quantum mechanics. In particular, **it is forever boiling with spontaneous fluctuations.** When we calculate these fluctuations we find that they grow, as a fraction of the distance, for nearby points. Eventually, for distances below about 10 e-33 cm., the calculated fluctuations in distance become larger than the distance itself. **Below this so-called "Planck length" our usual methods of calculation break down.** Indeed, the whole concept of distance comes to look suspect. Now 10 e-33 cm. is a very small distance, far beyond practical access. Nevertheless this issue is of fundamental interest, not only in its own right, but also for cosmology. **Indeed, our equations break down in describing extremely short <u>time</u> intervals** (10 e-44 sec.), for similar reasons. Thus we aren't able to describe the very earliest moments of the big bang. And so ultimate questions of origins remain up for grabs."*

According to the String-Brane Theory, what we know by **vacuum could be formed of a mesh (quantum foam) of micro-spaces (branes) of Calabi-Tau shapes of 6 dimensions** that would form the smallest particles of Our Universe. It is unknown to what scales these 6D-branes would appear, but they would be very close to the Planck dimension.

Then, for our usual methods of calculation (mainstream theories) Planck dimension seems to be the smallest dimension we can consider. But that does not mean that there (within Planck Dimension-Volume) cannot exist smaller things (objects, concepts, effects,...). It only mean that **our mainstream theories are not good enough,** and that they have to be enlarged to other theories that could **understand and parametrize all what happen inside the Planck Volume (below Planck Scale).**

Fig.15: 6D Calabi-Yau shape

If we consider the Calabi-Yau 6D-branes that, according to Brane-String theory , may exist at Planck Dimension-Volume, **we could consider these 6D-Branes as other micro-universes (similar to Our 3D Universe) that contain 6D Space,** with other (emergent) concepts (objects, effects, interactions,...) and Laws. **Quantic Fluctuations could be only effects coming from these 6D Calabi-Yau universes / worlds.**

To investigate what happens beyond the Planck size we need very high energies (above the Planck energy), and according to our current theories (QM and GR), this would create black holes. And if we increase this energy, **we could increase the size of**

these black holes.

This suggests that **our description of space-time is not right for these scales:**

- **Ed Witten:** *"Space and Time may have their days numbered"*
- **Nathan Seiberg:** *"I'm pretty sure that **space and time are only illusions"***
- **Nathan Seiberg:** *"**Space and time are likely to be emergent notions,** They are not present in the fundamental formulation of the theory, but appear as approximate macroscopic concepts." Emergent Spacetime, 2006 (http://arxiv.org/find/hep-th/1/au:+Seiberg_N/0/1/0/ all/0/1)*
- **David Gross:** *"Most likely, **space and time have** (elemental / fundamental) **components,** and they can be only emergent properties that arise in a theory with a very different look."*

Reaching a deeper understanding of the fundamental nature of space and time, it is one of the largest and intriguing challenges to physicists of next years.

A final consideration about space (empty space or vacuum) could be that, as it seems not to be so empty as we thought, and that space has life or consistency by itself, then, **might we consider space as the famous ether?.**

We know that **Gravitational waves are ripples in the curvature of spacetime** which propagate as waves, traveling outward from the source.

Might EM waves propagate also through this medium (the "substance" of empty space or vacuum) ?

What yes it seems to be possible is that **"empty" space might just be some unknown "substance"**, made up of much smaller components currently unknown (from a much smaller scale), that still should be determined and discovered. **This is what could give the empty space ("vacuum") a fractal (hierarchical) structure.**

ENERGY & SIZE SCALE RELATIONSHIP

Each space/size scale has associated an energy (mass) scale, (*"Warped Passages: Unraveling the Universe's Hidden Dimensions. Lisa Randall ,2005):*

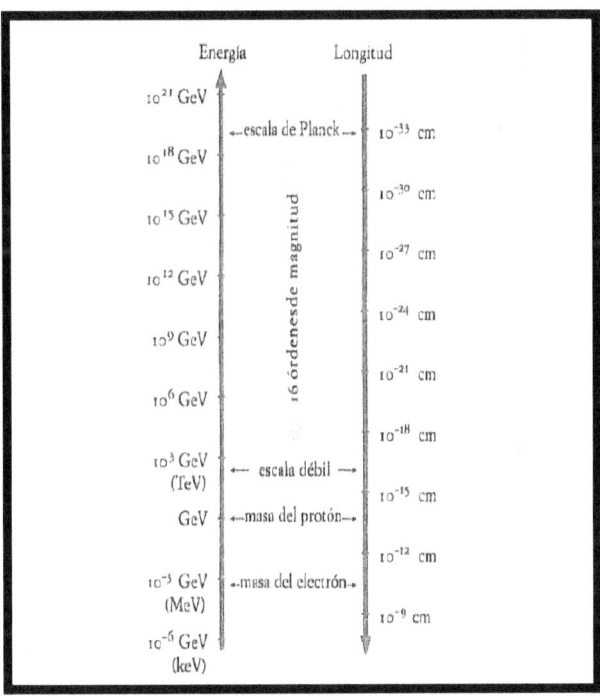

Fig.16: Relation between Energy & Length scales.
("Warped passages", Lisa Randall, 2005).

And its relationship can be described as:

> ***ENERGY** (wave-particle function) (GeV)   =*
> *K .[1 / **WAVELENGTH** (Distance, m)] **10 e +15***

From Uncertainty Principle (and also Einstein & De Broglie principles) **we can say that the different space scales (particles sizes or wavelength), are inversely related to wave energy (interaction) or particle mass** (quantum physics usually measure masses of particles in energy units: GeV).

Wave-particle duality: *Any wave has associated a particle (Einstein:* **EM -> photon**) *and any particle has associated a wave (de Broglie:* **particle -> Wave function**)

Einstein: E=h.v, where v is the wavelength and h the Planck constant. *E is the Energy of the EM particle (photon). Higher is the v higher is the E.*

Broglie: λ= h/mv λ= wavelength of a particle; mv = momentum (*= approx. Energy*)

As λ=1/v then we can say: E = m v

Energy Scales in Physics (*Source: Princeton University*)

*"**Energy scales decrease as length scales increase:** this is because of the* **quantum-mechanical relation** $\ell = hc/E$ *between a length scale* ℓ *and an energy scale E. Here h is Planck's constant and c is the speed of light. A useful approximate value is* **hc=200 MeVfm**. *One intuitive translation of the formula* $\ell = hc/E$ *is that* **a particle with rest energy E can't be confined to a region of space whose diameter is significantly less than** ℓ. *If you try, then there is enough uncertainty in energy and momentum to create* **particle-anti-particle pairs** *that tend to escape the spatial region to which you're trying to confine the original particle.*

This is one of the few existing relationships that is associated with the scale and for this reason I have considered it interesting to include it in the book.

This relationship between dimensional scale and energy **basically tells us two things:**

- **As the wavelength of an electromagnetic wave decreases, the energy of the wave increases,** reaching a minimum dimension (the Planck Dimension) in which the energy of the wave is the maximum allowed by our laws (the Energy of Planck), although this is not infinite.

- To know what happens to **very small scales it is necessary to use collisions (CERN-LHC) with increasing collision energies.**

It is common in Quantum (Particle) Physics to express mass in units of Energy (eV/c²):

- **Proton mass:** 1,78. 10 e-27 Kg = **1 GeV**/c²

Due to Einstein relation: **E = m·c².**

DSR THEORY (HIGH ENERGY SCALE)

In the same way that the theories of MOND (see chapter 5) are proposing alternatives to the laws of Newton and GR for very high spatial scales (> 10 e+20 m), there are also other QG theories proposing alternatives for very small (or very high energy) spatial scales.

For example the **DSR (= Doubly or Deformed SR)** theory which **proposes that SR is not valid for High Energies** (Planck scale), and it forecast that light speed could increase till infinite for Planck Energy (c = f (E))

Doubly special relativity (**DSR**) – also called **deformed special relativity** or, by some, **extra-special relativity** – is a **modified theory of special relativity** in which **there is not only an observer-independent maximum velocity** (the speed of light), but an observer-independent **maximum energy scale and minimum length scale** (the Planck energy and Planck length).

SR is based on two principles: (1) the relativity of motion , and (2) the invariance and universality of the speed of light.

DSR I assumes that the Planck dimension is the smallest object that can be seen, and its dimension is the same for all observers (whether they are stopped or in motion), as it is for speed of light. **Both would be two universal variables (speed of light and Planck length).**

DSR I theory also assumes that the Planck energy is the maximum energy that one elementary particle can have. Currently the maximum detected energy is 10 e -9 times this maximum Planck energy (by cosmic ray detector AGASA).

Further **DSR II** assumed that **at very high energies the speed of light increases to infinity at the Planck energy**.

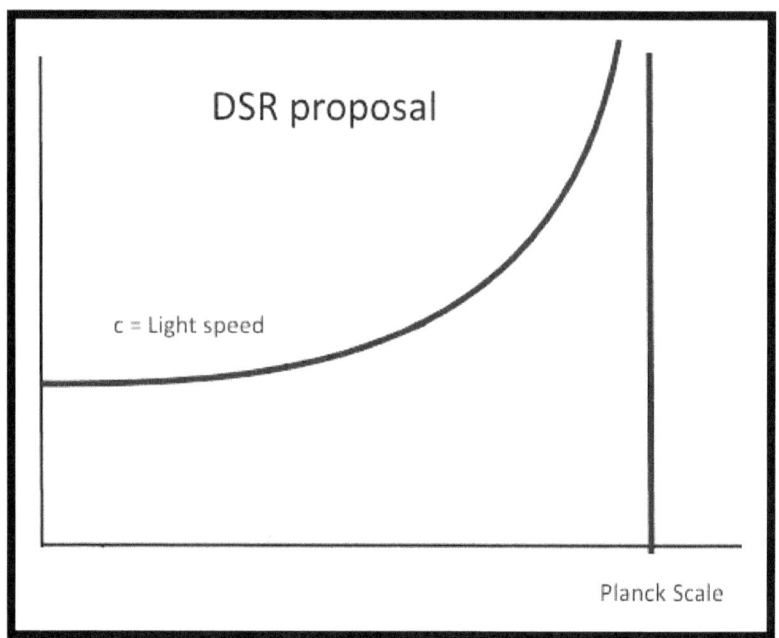

Fig.17: DSR Diagram [c= F(E)]

UNCERTAINTY PRINCIPLE

Stephen Hawking says that quantum mechanics itself is deterministic, and it is possible that the apparent indeterminacy really is because **there are no particle positions and velocities, but only waves.**

So for Stephen Hawking, *"uncertainty principle is only apparent, but not real".* There could be another way of seeing the Universe, in another scale, where concepts and laws are different.

*In quantum mechanics, the **Heisenberg's uncertainty principle**, is any of a variety of mathematical inequalities asserting a fundamental limit to the precision with which certain pairs of physical properties of a particle, known as complementary variables, such as position x and momentum p, can be known simultaneously.*

*Introduced first in 1927, by the German physicist Werner Heisenberg, it states that the **more precisely the position of some particle is determined, the less precisely its momentum can be known, and vice versa.** The formal inequality relating*

Uncertainty Principle has been always a strange and unintelligible concept. But mainly it **could be a problem of measuring instruments**, because we try to measure them (location and momentum) precisely with inappropriate instruments (larger than required). **It is not appropriate to use EM waves (photons) to measure the position and speed of an electron.**

Surely if **we could see the uncertainty principle from his own quantum scale**, we could see that we can determine the velocity and position of a particle (electron) at the same time. But we should detect it by other means, and not with those who we know nowadays (photons). **Possibly we could use "waves" of nuclear interactions (strong and weak)?** Or other not wave-based sources ?.This, provided that the concepts position and velocity of a "particle" have some sense for these scales.

"As well known, Electromagnetic wave (EM wave) is caused by electromagnetic interaction, and gravitational wave (G wave) is caused by gravity; **we could propose that strong interaction wave (S wave) is caused by strong interaction, and weak interaction wave (W wave) is caused by weak interaction.**". From article "Fifteen Kinds of Waves Caused by Four Fundamental Forces", **Fu Yuhua, Fu Anjie, Zhao Ge (Beijing Relativity Theory Research Federation).**

What will happen with HUP if that proposal/assertion is true?

Mainstream scientist consider that the uncertainty principle is not only a measurement limitation, but also more fundamental than that. As far as we know by now, it simply **doesn't matter how we take the measurement, we will never improve on its limits**.

But we also must consider the option that **Uncertainty Principle could be simply a matter of trying to understand phenomena, typical of other scalar spectra, with the parameters and models of our own scale spectrum.**

If we agree that, on a quantum scale, everything behaves like waves, and what for us it is a particle (matter), it is just one particular type of wave (as stated in the String Theory), then **concepts such as position and speed (momentum) do not have the same meaning as they have for our own scale**. We are simply trying to understand other phenomena from other spatial scale, with (emerging) concepts of our own spatial scale.

Violation of Heisenberg's Measurement-Disturbance Relationship by Weak Measurements (Lee A. Rozema, Ardavan Darabi, Dylan H. Mahler, Alex Hayat, Yasaman Soudagar, and Aephraim M. Steinberg Phys. Rev. Lett. 109, 100404 – Published 6 September 2012; Erratum Phys. Rev. Lett. 109, 189902 (2012):

"While there is a rigorously proven relationship about uncertainties intrinsic to any quantum system, often referred to as "Heisenberg's uncertainty principle, Heisenberg originally formulated his ideas in terms of a relationship between the precision of a measurement and the disturbance it must create. Although this latter relationship is not rigorously proven, it is commonly believed (and taught) as an aspect of the broader uncertainty principle. Here, we experimentally observe a violation of Heisenberg's "measurement-disturbance relationship", using weak measurements to characterize a quantum system before and after it interacts with a measurement apparatus. Our experiment implements a 2010 proposal of Lund and Wiseman to confirm a revised measurement-disturbance relationship derived by Ozawa in 2003. Its results have broad implications for the foundations of quantum mechanics and for practical issues in quantum measurement."

THE NUCLEAR WAVES

If there are electromagnetic waves and gravitational waves, shouldn't be there weak nuclear and strong nuclear waves?

Although it is not obvious or imperative, it seems logical and possible that may exist **Nuclear Waves** produced by the nuclear force-fields (strong and weak), **in the same way** that the **electromagnetic and gravitational** forces fields have their own respective waves.

If these **Nuclear Waves** exist, obviously they would have a very short range (scope), and possibly, therefore, we have not yet detected them. (See different opinions about at the end of this introduction):

Here there are some answers to this question with **different opinions:**

Yes, there are and we have observed them:

- For the **weak force the waves are called W and Z bosons,** they have rest energy (mass) and charge too. They are the mediator of the weak force.
- For the **strong force the waves are called gluons.** They do not interact with electric fields. They have this threefold property called color. They mediate the strong nuclear force.

Electromagnetic wave is also a particle and we call it the photon, Or we could say that the photon is the carrier particle of the electromagnetic force.

In the same way, the **W and Z bozons are** the carriers of the strong force, and you can say that they are the **"weak wave"** in the same way that the photon is electromagnetic wave. However, these bosons are very massive and consequently, **have a very short life before they decay**. So these **"weak waves" don't get a chance to travel very far before they get transformed into something else.** Of these bosons, the Z is, in many respects, just like a heavy photon; the W+/W-, however, also carry electric charge.

The same goes for gluons and the strong force. **Gluons are the carriers of the strong force.** This particle is **massless**, but because the nature of the strong force is such that it actually increases with separa-

---

tion, **free gluons exist only in very high energy environments**, such as inside a particle accelerator, or in the extreme early universe. Unlike the photon, gluons are not neutral; they carry various combinations of the QCD "color" charge.

As for gravity, the detection of gravitational wave is just more evidence for the graviton particle.

https://www.quora.com/If-there-are-electromagnetic-waves-and-gravitational-waves-shouldnt-there-be-weak-nuclear-and-strong-nuclear-waves

---

No, there are no weak or strong waves in the sense as there are for electromagnetic or gravitational waves.

*The electromagnetic and gravitational waves are classical objects, they are possible vacuum solutions to the classical equation of motion for the field strength of the respective force, and can be radiated by objects charged under the respective force. But **the weak and the strong force have no analogous classical limits - the weak force is suppressed by a factor***

*Due to the mass of the W- and Z-bosons and thus very different from the EM or gravitational force, and **it doesn't make sense to speak of a classical limit of the strong force because gluons and quarks are confined** - there are no net charges under the strong force on a classical level, hence the strong force just vanished from our description.*

In other words, the weak and the strong force are, in some sense, "fully quantum" *in that their importance to our world comes completely from their quantized description, and a classical description does not make sense for them, thus **we cannot speak of a classical concept such as a wave for the weak and strong forces.***

http://physics.stackexchange.com/questions/223424/are-there-weak-force-waves

---

1.- There can't be "strong waves" or "weak waves". *Just like EM waves are coherent states of photons, and gravitational waves are coherent states of gravitons, **the strong waves or weak waves would be coherent states of gluons or W/Z bosons.***

*However, **gluons don't really exist**. That is, there is a **gluon field, but it cannot have waves**, and you really cannot create a state with a gluon. This has to do with a property of Yang-Mills theories (of which quantum chromodynamics is an example), namely that they have a nega-*

tive beta function - rougly, the higher the energy scale, the smaller the strength of the force. (EM does exactly the opposite).

This means two obvious things: 1) in the limit of very high energies, the colour force becomes weak. This is **asymptotic freedom**. 2) in the limit of very low energies (such as room temperature) the colour force is extremely strong. This is **confinement**.

Confinement is what is important to us. This effect makes sure the force between coloured states is so large that they cannot ever be seen. Taking a meson and trying to pull the quark and antiquark apart does not allow you to see the bare colour of the quarks - you just get more mesons.

So confinement is commonly formulated at: there exist only colourless, or white, states.

Indeed **hadronic physics has a very rich spectrum of thousands of states bound by the strong force**, baryons and mesons (and more exotic beasts apparently), **all rigorously white.**

Now in Yang-Mills theory (and so also in QCD) the gauge bosons themselves (the carrier particles) are also charged under the interaction they mediate. (Precisely, they're in the adjoint representation). So if there was such a thing as an actual gluon, which is coloured, you could change reference frame to get it to an arbitrarily low energy (you can do this only because the gluon is massless). **As the energy is very low, confinement applies and you get a contradiction.**

More practically gluons interact with other virtual gluons constantly, in such a strong way that a clean state with a single gluon does not simply exist, **let alone coherent states of gluons in macroscopic waves. It just cannot be created.** If you give energy to the vacuum, all you'll be able to do is create colourless states. No gluons, and no quarks either. The lightest particle possible in QCD (the mass gap) is the pion.

People often say the colour force is short-ranged, and so this would be a very easy argument against colour waves. This is incorrect, since the range of the strong force is infinite as the gluon is massless. Instead, the actual reason is QCD is confining, and all of the above.

For the weak force, the mediating bosons are very massive and very unstable. W bosons decay to quark-antiquark or lepton-antineutrino, and they do so in the general order of 10^{-25} seconds. So that immediately kills the idea of a "weak wave".

2.- While we can't detect "strong force waves" and "weak force waves" in the same sense we can detect gravity and electromagnetic waves, **we**

can detect ripples in the strong force field and the weak force field that are analogous to photons; *these are the gluons and the W/Z bosons. Both gluons and W/Z bosons are very short-lived, but nonetheless we can detect them by their decay products. I think goes a bit too far in implying there is no such thing as a gluon.* ***We can "see" high energy gluons in high energy collisions at particle accelerators:*** *each produces a "jet", ie a "shower" of particles. But it's true that* ***the ripple associated with a high energy gluon very quickly disintegrates into many thousands of ripples*** *in both quark and gluon fields, until you end up with relatively stable bound states like pions, neutrons, etc.*

https://www.reddit.com/r/askscience/comments/45farb/could_we_detect_strong_force_waves_and_weak_force/

7. EMERGENT CONCEPTS & LAWS

Our vision (and knowledge) of the different scale LANDSCAPES is always from our own scale spectra (our LANDSCAPE: Newtonian Landscape). And from there we try to understand everything that happens in the others scale landscapes.

But, what would happen if we could observe these same LANDSCAPES from their own scale?

It could give us a completely different point of view that we have now, and **it would surely help us to understand better many concepts and phenomena that now we do not fully understand.**

We could consider that **most Physical concepts** (such as vacuum, energy, matter, space, time, speed,...), **and also most physical theories or laws** (such as Newtown, Maxwell, Thermodynamics, Relativity, Quantum,...), they **are just emergent concepts and theories**. This proposal is based on *__Robert B. Laughlin__* (*"Different Universe: Reinventing Physics from the Bottom Down"*, *2006*). *See Annex 2.*

Concepts so usual for us (for our LANDSCAPE) such as energy (matter), vacuum, and time may have no meaning for another LANDSCAPE. In the same way that thermodynamic concepts such as temperature and pressure are values (emerging) that only make sense for a larger size of a atom (> 10 and -15 m) and they are meaningless to smaller dimensions. **We can say that energy (matter), vacuum and time are emergent concepts.**

A Different Universe: Reinventing Physics from the Bottom Down, Robert. B. Laughlin, 2006.

In this age of superstring theories and Big Bang cosmology, we're used to thinking of the unknown as impossibly distant from our everyday lives. But in A Different Universe, Nobel Laureate Robert Laughlin argues that the scientific frontier is right under our fingers. Instead of looking for ultimate theories, Laughlin considers the world of emergent properties-meaning the properties, such as the hardness and shape of a crystal, that result from the organization of large numbers of atoms. Laughlin shows us how the most fundamental laws of physics are in fact emergent. A Different Universe is a truly mind-bending book that shows us why everything we think about fundamental physical laws needs to change.

POINT OF VIEW (LANDSCAPES)

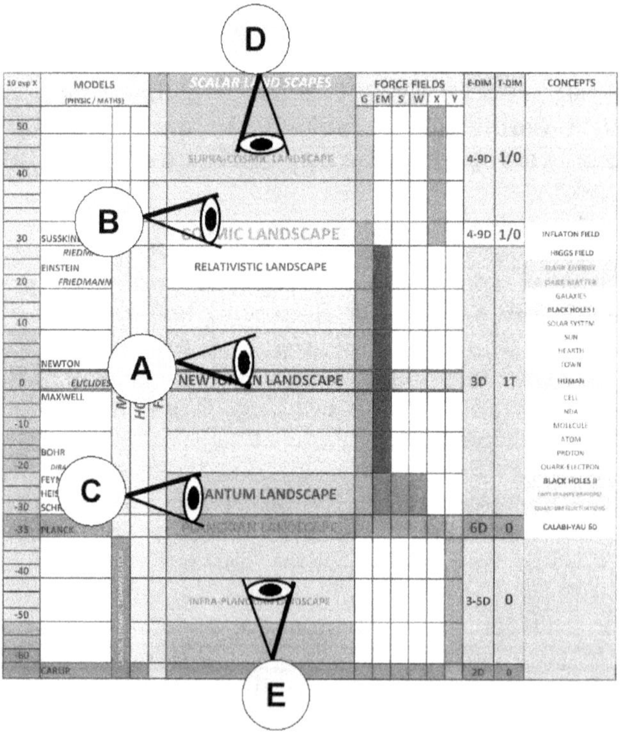

Fig.18: Point of View from different Landscapes

For each LANDSCAPE there are some physical models that best explain their behaviors: Newton, Maxwell, Chemistry, QED, QCD, Einstein (SR-GR) . Although they all may be related to each other by some underlying laws, we **can also consider that the different laws / models are emergent and depending on the scale (Landscape).**

Current effort on ToE is finding these underlying laws that can unify all models, and especially the SR-GR with QM (QED, QCD & QG). But, at best, these **ToE (String-Brane Theory) only will be able to explain a wider spectrum**, which includes several LANDSCAPES (between Our Known Universe scale, 10 e +27 m, to Planck scale 10 e-35 m). It is only a way of widening the scale spectra. **And, possible, other further ToEs could wide more this spectra.**
The laws of physics are the same throughout the whole space, but it does **not mean** that **the scenario in which they are wrapped will be always the same** (spaces of different dimensions or different size/space scales), **where different laws may emerge.** Although all of these emerging laws may be governed by some basic underlying laws.

Einstein himself never considered the Relativity Theory as a fundamental theory, and he expected improvements on it due to Quantum Physics.

EMERGENT INTERACTION FIELDS

As we have mentioned previously the known interaton fields are: Gravity, EM and Nuclear fields (strong and weak).

Electroweak Theory (S.Glasgow, S.Weingberg & A. Salam, 1963-67), propose that weak and EM interactions arise from the same **electroweak interaction** at very high energies.

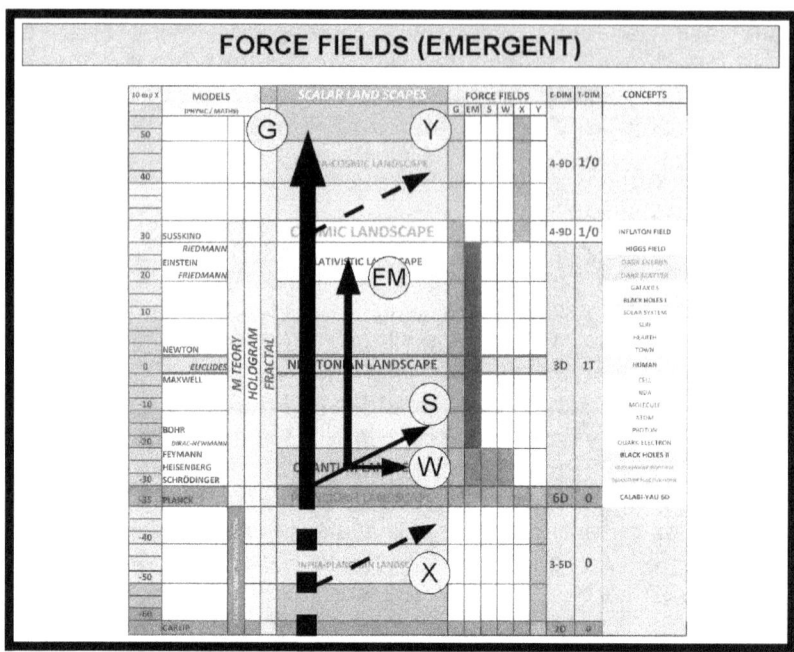

Fig.19: Emergent Interaction Fields

And **Supersymmetric Theory** forecast that **EM, S & W interactions** split off from a single, unified force due to spontaneous symmetry breaking at the very early universe. (prior to 10-11 seconds after the beginning). Therefore, **at very high energies (very small spatial scales) would be the same interaction (force).**

Unified Field Theory (**UFT**), is a type of field theory that allows all that is usually thought of as fundamental forces and elementary particles to be written in terms of a single field. **There is no accepted unified field theory, and thus it remains an open line of research.**

Gravity Field refuses to be unified, and **seems to follow different principles.**

May we also consider that known interactions (Gravity, EM, Weak and Strong) as emerging effects ?

What would happen if we could see/observe (detect / measure) within a very small volume (< 10 e-25 m or smaller than Planck volume)?, there will be also 2-3 interactions ?, or only one (unified), ... or none ?

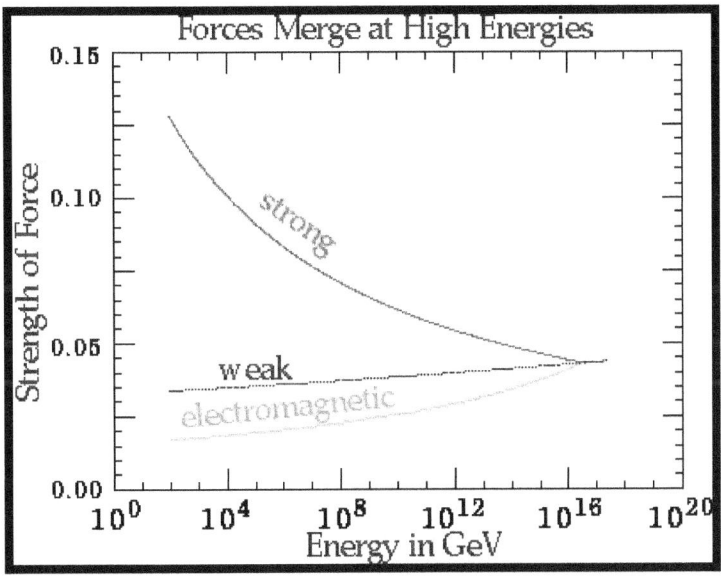

Fig.20: Interaction field unification

Coupling Constants for the Fundamental Forces:

In attributing a relative strength to the four fundamental forces, it has been proved useful to quote the strength in terms of a coupling constant. **The coupling constant for each force is a dimensionless constant.**

Coupling Constants (values):

Strong	$\alpha s = 1$
Electromagnetic	$\alpha e = 1/137$ (= Fine-structure constant)
Weak	$\alpha w = 10\ e\text{-}6$
Gravity	$\alpha g = 10\ e\text{-}39$

*Otherwise, we can also say that the **different interactions (forces)   fields have different scope of action**:*

Scope of Fg = infinite *(Whole Universe or "Bulk". Due that Gravity is a closed string and it can go out of Our D-Brane, to the Bulk)*
Scope of Fem < 10 e +27 m *(Within Our Universe radius. Due that EM is an open string and cannot go out of Our D-Brane)*
Scope of Fs < 10 e -15 m *(Atom -Nuclear radius)*
Scope of Fw < 10 e -18 m *(Quark radius)*

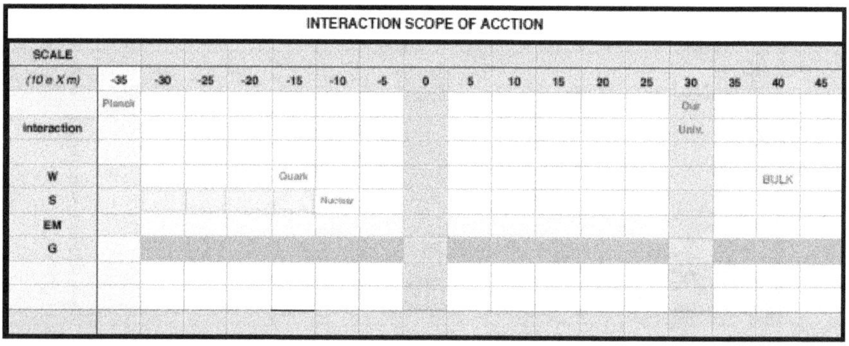

INTERACTION SCOPE OF ACCTION																
SCALE																
(10 e X m)	-35	-30	-25	-20	-15	-10	-5	0	5	10	15	20	25	30	35	40	45
Interaction	Planck														Our Univ.		
W					Quark										BULK		
S						Nuclear											
EM																	
G																	

Fig.21: Scope of action of the Interaction Field

Three of these interactions (**EM, Weak and Strong**) seem to be unified to very small size (large energy) scales (approx. Planck Dimension/Energy), as if **they arose from a single force** from there (possibly from the 6D shapes: Calabi-Yau?).

On the other hand, **the Gravitational interaction seems to follow different patterns.** And, as Lisa Randall propose, the "*mass seems to appear as if by magic, and it is 10 e-16 times (order of magnitude) weaker than physicists could expect only from general theoretical bases (the **Hierarchy Problem**)*".

Moreover, for larger size scales (> 10 e+30 m), if we could go out of our 4D universe, and we could see it from outside (from the Cosmic Landscape), possibly we could detect the Gravity interaction, but also **we could detect other emergent interactions, currently unknown.**

The **Force of Gravity itself is a force clearly emergent** and difficult to foresee for someone who only knew the laws governing small scales (within an atom, < 10 e -20 m), since it begins to have importance when grouping many Atoms and molecules (possibly we cannot clearly detect their influence before having the mass of a rock over 100 km in length). In the same way an **Emerging Force (Y) can appear for very large scales** (In the Supra-relativistic or Cosmic Landscape,> 10 e +20 m).

If this were so, **these new forces/interactions could also be another alternative explanation to the Dark Matter,** they could emerge on these scales due to the clustering of stars and/or galaxies, generating unintended effects and behaviors.

Following I relate a very short and relevant text **(Markus Hanke, The Science Forum, April.2015)** that I agree with and I consider very clever and fascinating:

*"In the Standard Model as it currently stands, there are 25+ quantum fields (of which the electromagnetic field is only one), all of which **extend throughout the universe**, i.e. they are present everywhere; their mere presence does not imply that gravity is somehow "caused" by any of them, and less still does it single out electromagnetism in any way. Several people, including myself, have pointed out already that **electromagnetism behaves very differently from gravity in many fundamental respects. T**he two are quite simply, not the same thing, and neither one of which causes the other Although, of course, they influence one another since electromagnetic fields carry energy-momentum. However, what **we can do is unify gravity and electromagnetism into a common framework** - this is done **by simply adding a spatial dimension to the universe**, which is curled up into a small "circle" at each point of space-time, meaning it isn't seen on macro-scales. In essence, it turns out that **5-dimensional General Relativity is exactly equivalent to 4-dimensional GR + 4-dimensional electromagnetism, plus an additional scalar field**; this model of a <u>5-dimensional universe is called Kaluza-Klein gravity.</u> The implication of this is that <u>**gravity and electromagnetism, though different by nature, both arise from the same underlying mechanism,**</u> being the geometry of a space-time with a compactified fifth spatial dimension; gravity and electromagnetism are, under this model, both geometric properties of space-time, and hence on equal footing (but not the same thing !). Unfortunately though Kaluza-Klein gravity implies the existence of an additional scalar field (often called the **<u>"dilation field"</u>**), for which there is no empirical evidence whatsoever - which is why this model never made it to the mainstream, since the dilation should have been easily detectable even for first generation particle accelerators, but it just isn't there. Nonetheless, **it is a***

fascinating model and definitely of academic interest."

It has been shown that **EM and Nuclear Interactions (strong and weak) are unified to very high energy scales** (near Planck energies and Planck scale). Electroweak theory unified both EM and weak (Steven Weinberg, 1993: "The Search for the Fundamental Laws of Nature").   And **according to string theory (ToE) it is also believed that gravity can be unified with them, to higher energy levels.**

The laws of these forces (EM-S-W-G) are somehow related (albeit in different spatial dimensions), so we can say that **this known forces/interactions are simply different manifestations of the same force/interaction for different spatial dimensions.**

---

*"**There may exist other worlds** that we do not know, in other branes separated from ours by other hidden dimensions. And, **if there is life** on some of these other branes, it is likely that these beings will be trapped in a completely different environment. And **they should feel different forces/waves that would be detected by different senses**" (Lisa Randall).*

KALUZA-KLEIN THEORY

*In physics, **Kaluza–Klein theory (KK theory)** is a **unified field theory of gravitation and electromagnetism built around the idea of a fifth dimension beyond the usual four of space and time.** It is considered to be an important precursor to string theory.*

* *The original hypothesis came from Theodor Kaluza, who sent his results to Einstein in 1919, and published them in 1921. **Kaluza's theory was a purely classical extension of general relativity to five dimensions.** The **5-dimensional Einstein equations yield** the 4-dimensional Einstein field equations, the Maxwell equations for the electromagnetic field, and an equation for the scalar field.*
* *In 1926, **Oskar Klein gave Kaluza's classical 5-dimensional theory a quantum interpretation**, to accord with the then recent discoveries of Heisenberg and Schrödinger. Klein introduced the hypothesis that the **fifth dimension was curled up and microscopic**, to explain the cylinder condition. Klein also calculated a scale for the fifth dimension based on the quantum of charge.*

*Currently, on the original idea of Kaluza and Klein it has been constructed generalizations of **the theory of relativity of space-time of more than five dimensions**. These theories are also usually called theories of Kaluza-Klein*

TIME CONCEPT

According to Lee Smolin *"for the future development of physical science there is something that escapes us, and among these concepts is the Nature of Time"*.

Before Einstein, **Time was considered an independent and absolute concept/dimension,** and always **with arrow positive** (from the past to the future) and constant/homogeneous (moving at the same rate/speed).

SR (Einstein's) theory proposes that the **Time is variable depending on the speed** of the object (the higher is the speed of an object, time will run slower for him, relatively to another object moving at a lower speed). And also, **Time varies according the gravity force that is exposed to the object** (the higher is the gravitational force on an object, then slower will run the time relatively to another object that is exposed to a lower force). **If an object is exposed to a infinity Gravity (Black Hole) Time will be zero.**

Also Time is null (zero) for particles moving to the light speed (EM radiation,...). Even more, **whether information** (object without mass) **that move at a speed faster than light** (which is now considered impossible), its **time would be negative**, and therefore, such information **would travel to the past (!?)**.

Stephen Hawking (*"A Brief History of Time", 1988*) proposed that **the arrow of Time** could **depend on the entropy (S),** being positive if entropy rate is positive, and, (possibly) negative, if entropy rate is negative (which would contradict the Second Law of Thermodynamics). If this proposal is OK the arrow of **Time could be different for different "pocket" universes** depending on their entropy. Realize that **entropy is always positive (S>0)** but we are talking about **entropy rate** (entropy variation $\triangle$**S**).

Entropy (S) is a thermodynamic concept that measures the disorder of a physical system. And the <u>*Second Law of Thermodynamics*</u> *states that the current S of the Universe increases (* $\triangle$ *S is positive).*

While Our Universe is expanding, entropy is increasing , and time is positive. But what will happen if Our Universe (or other "pocket" universe) implode ("**Big-Crunch**")? S.Hawking, initialy supose that time could be negative, but finally accepted that **during the**

implosion (although the entropy rate could be negative), **Time will be also positive.**

> *"**Thomas Gold, in 1958** (before arrival of Quantum Cosmology) was the first to propose that **during the "Big Crunch" time could be negative**. This behavior of the volumen is quite possible according to the theory of General Relativity. Currently we have to admit that there is a remote possibility that this behavior occurs because **the nature of time in Quantum Gravity is not yet entirely clear.**"*
>
> *"However, **it is much more likely that such reverse is simply an illusion** generated by the mathematical description chosen. When we walks on the earth on a path that crosses the North Pole, latitude increases initially and, decrease after crossing the Pole. But that does not mean it has returned to previous path as the degrees of longitude are different. Similarly, **the reversal point of the volume of the universe as a parameter of time in a collapse ("Big Crunch"), does not mean that the time elapsed in reverse (be negative).**"*
>
> *(M. Bojowald, "Before the Big-bang", 2009)*

We could say that if Time arrow is proportional to the Entropy rate ($T = f(\triangle S)$), then **Time arrow rate could be variable depending on the Entropy rate value at different ages of Our Universe since Big-bang.**

For all the above, we could extrapolate (*"The Fabric of the Cosmos: Space, Time, and the Texture of Reality"*, 2005 by Brian Greene) that **Time began for Our Universe with the Big Bang**, and that it has, to date, evolved with a positive arrow, but, probably, to different rates depending on the value of the entropy on every moment (depending on the entropy rate for every ages of Our Universe).

If we go further, and **we settle beyond the scalar limits of Our Universe (on the Cosmic Landscape),** with all their different bubble (pocket) universes (with different constants and evolution state), we can extrapolate that **the Arrow of Time for each universe could have a different value/rate, depending on its own characteristics** (Entropy, gravity, speed of light,...), and possibly, depending on its state of evolution (Expansion-Implosion).

We may also assume what the **Time in the D-space inter-verses of the Cosmic Landscape** (space between different bubble-pocket verses), **should be also proportional to the entropy,** but should also depend on **other** known **variables** (gravity, speed of light,...), and possibly other variables presently unknown.

May we consider Time as an Emergent concept ?. If Time depends on Entropy, we could ask if we could consider entropy on the Quantum-Planck scale spectra (?). Entropy is a measurement of the disorder of the universe (particle-wave, mass-energy). Is there sense for entropy concept below 10 e-20 m scale ? **Is there Time below this 10 e-20 m scale ?**

In quantum gravity, may be there is no notion of absolute time. Like all other quantities in the QG theory, the notion of time has to be introduced "relationally", by studying the behavior of some physical quantities in terms of others chosen as a "clock".

QUANTUM TIME

*While time is a continuous quantity in both standard quantum mechanics and general relativity, many physicists have suggested that a **discrete model of time might work better**, especially when considering the combination of quantum mechanics with general relativity to produce a **theory of quantum gravity**.*
*A **chronon** is a proposed quantum of time, that is, a discrete and indivisible "unit" of time as part of a hypothesis that proposes that time is not continuous.*

Current theoretical physics suggests the flow of time is just an illusion.

- *"The Time concept appears when processes of change or motion occur."*
- *"**In a stationary Universe (with no changes or not movements), the time will not exist.**"*
- *"It is not true that we move through space, **we move through space-time**. The space and time are inextricably linked."*
- *"The nature of Time in Quantum Gravitation is not properly explained".*
- *"**Time arrow cannot be attributed to the entropy at microscopic scales**".*
  (M. Boljoland, "Before the Big-bang", 2009)

---

http://fqxi.org/community/essay/winners/2008.1
The Nature of Time by Julian Barbour: *"A review of some basic facts of classical dynamics shows that time, or precisely duration, is redundant as a fundamental concept. **Duration and the behaviour of clocks emerge from a timeless law that governs change.**"*
Does Time Exist in Quantum Gravity? *by Claus Kiefer: "Time is absolute in standard quantum theory and in general relativity. [...] Among the consequences are the fundamental timelessness of quantum gravity, the approximate nature of a semiclassical time, and the correlation of entropy with the size of the Universe."*

TRAVEL IN TIME

According to special relativity (Einstein), when a body A travels at very high speed V (a ratio of the speed of light: 10, 20,..., 80%) compared to the other that stays in resting B, time elapses for A slower than for B inversely proportional to the speed:

If $V = c \cdot 50\%$ (50% the speed of light (c) = 150,000 km/s)
$Ta = Tb \cdot (c-V)/c = Tb/2$ (if Tb =1 year, then Ta= 2 years).
Then if A and B are in the same place at given instant To, and A initiates a trip to the star Alpha Centauri (4 light years far from our Sun), at an average rate of 50% the speed of light (so for A time will run half as fast as for B that rest in hearth), the round trip for A will be 8 years, while for B (rest in hearth) will be 16 years, when A return to Earth.

Therefore **travel to the future are possible, as long as we can travel at very high speed relative to other bodies.**

Travel to the past, they are different, and they seem more improbable (and I would say impossible): As S. Hawking said, **"I still have not seen any tourist visiting us from the future."**

If **A could travel at the speed of light** (300,000 km /sec), time for A not passes. **Time for A would stop**.

And become the paradox that **if A was able to travel at a higher speed than the speed of light** (impossible theoretically), then the arrow of time would reverse, and **A would travel to the past.**

Another "theoretically" way to "travel" (to "watch") **to the past would be through a Thorne´s "wormhole".** Suppose A and B, where A and B have a "device" (possible "time machine") mobile screens (TV) connected by a "wormhole" through which they can see each other . If A device worn on a trip, A could still see B during the trip (passing equal time for both), and when A to return to Earth, A would continue seeing B by the screen (which for both passed the same time), but also live (where B would be older). If A had some way to travel through the "wormhole" (the device), then he could travel back in time. And if A can pass through the wormhole, anyone else could do it, and so in one direction (to the past) and to the other (the future). Then **we would have a "time machine".** But **the question would be, could we really "interact" with the past, or we could just observe it?** (Read "The Fabric of the Cosmos" by Brian Greene).

Obviously, **this option would only allow us to "travel" to the past, but only until the time we built this device (the "time machine"), but we could never travel to earlier times.**

8. EVENT HORIZONS

In general relativity, an **event horizon** is a boundary in spacetime beyond which events cannot affect an outside observer.

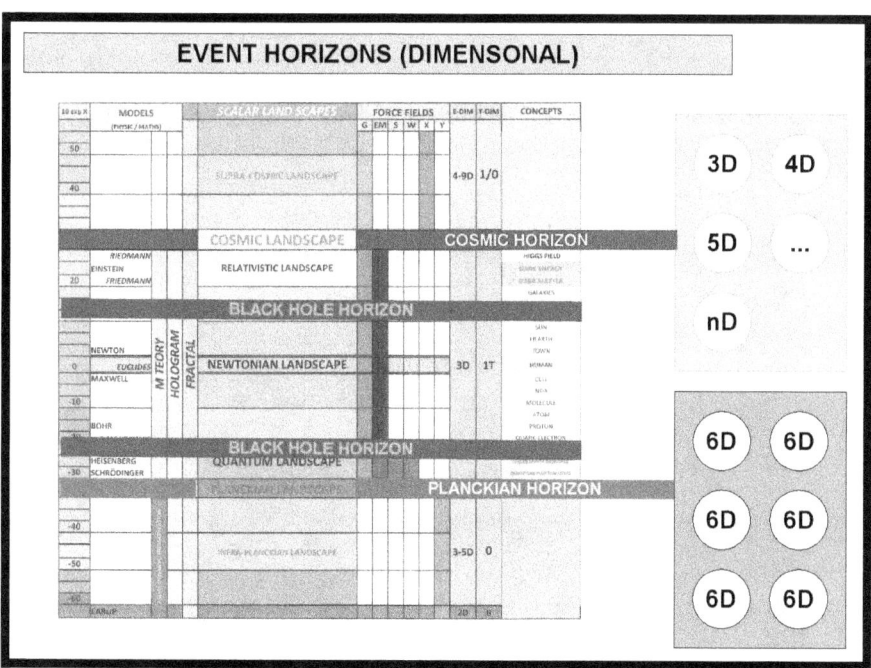

Fig.22: Universe Event Horizons

It is clear what it means the **"event horizon" of the Black Holes** (the external surface of the black hole) and its **"Hawking radiation".**

A **black hole** is a geometrically defined region of spacetime exhibiting such strong gravitational effects that nothing—including particles and electromagnetic radiation such as light—can escape from inside it.

S.Hawking proposed that **all the 3D info (entropy) contained inside a Black Hole,** can be described also on the 2D boundary of the Black Hole, and it **is equal to the Planck area units (qubits) of this surface:**

"The entropy (information) of a black hole is proportional to its surface area. Thus the amount of information required to specify the micro-state of the black hole is proportional to its surface area. Black holes have been shown, theoretically, to be maximally entropic and by putting together the surface area and the entropy set by the Beckstein bound you get the result **that each bit could be regarded as encoded in a Planck area.** Qubits require an entangled state and it is not sure proven that entanglement survives at this boundary"

Fig.23: Black Hole Horizon

> *"**Hawking Radiation** is produced due to quantum fluctuations (particle-antiparticle pairs) on the surface of the black hole, and being one of the particles trapped by the black hole, and the other escaping as radiation.*
>
> *<u>Large black holes </u>emit **Hawking radiation so dim** that temperature is less than the Cosmic Microwave Background radiation (CMB, 3º K), and this **prevents us from seeing and detect them.***
>
> ***Smaller black holes emit more energy and radiation,** exceeding CMB. Thus it is lost more energy than it is gained by absorption of background radiation, being able that black hole could be evaporated."*
>
> *("Before the Big-bang", Martin Bojowald,2009)*

In a similar way, **we could also consider other types of "Event Horizons" and, possibly, its own " X-radiation":**

- **<u>Our Cosmic Horizon (approx. 10 e+30 m)</u>**: The event horizon (boundary) between Our "pocket" Universe (Our 4D-Brane-world) and beyond of Our Universe (to the Bulk or Cosmic Landscape). As this boundary is further than the Observable Universe boundary, then **we will not be able to see or detect any signal from it .**

- **<u>Observable Cosmic Horizon (approx. 10 e+27 m):</u>** The event horizon (boundary) of the Observable Universe (That part of our universe that, due to the limitation of the speed of light, we are able to detect or observe). Possible cosmic microwave background **(CMB) could be also considered radiations of this horizon** (radiations coming from beyond Our Observable Universe boundary).

- **<u>Planck Horizon (< 10 e - 35 m):</u>** The possible event horizon (boundary or edge) where might end Our Universe spectra on the lower scales, or where it make the change (if boundary) between Our Universe spectra (Our 4D-Brane) with the Sub-Planck Landscape (possibly within 6D space Calabi-Yau branes-universes). **Quantum Fluctuations** (and possibly also Interaction Fields) **could be considered as effects ("radiations" ?) arising from this horizon** (from the Planck Volume or from the 6D Calabi-Yau shapes).

Fig.24: Observable Universe Horizon

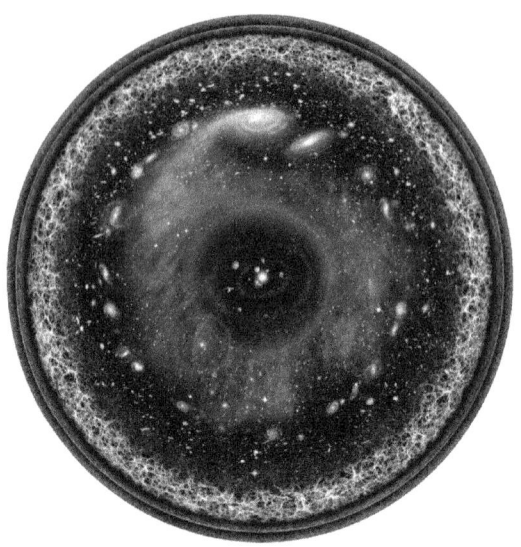

The first (**Our Cosmic**) and third (**Planck**) **horizons,** could be considered as **branes boundaries of Our Universe Brane** (Our Universe Scale Spectra) and other neighbors branes (Up: the Bulk, or Down: the Calabi-Yau).

*We could say that **Our Universe (Our Brane-World 3D) will be limited up by Our Cosmic Horizon and down by Planck Horizon**, that would separate it from other higher dimensional branes.*

This would be like a fish that lives in the sea, and is limited above by the surface of the sea that separates it from gaseous air. And down would be limited by the solid state of the seabed.

The different dimensional branes could be considered as different (dimensional) space states.

The second (**Observable Cosmic**) **horizon** can be considered only as a **physical limitation** due to the speed of light limitation.

> *"The CMB is a **cosmic background radiation** that is fundamental to observational cosmology because it **is the oldest light in the universe**, dating to the epoch of recombination."*

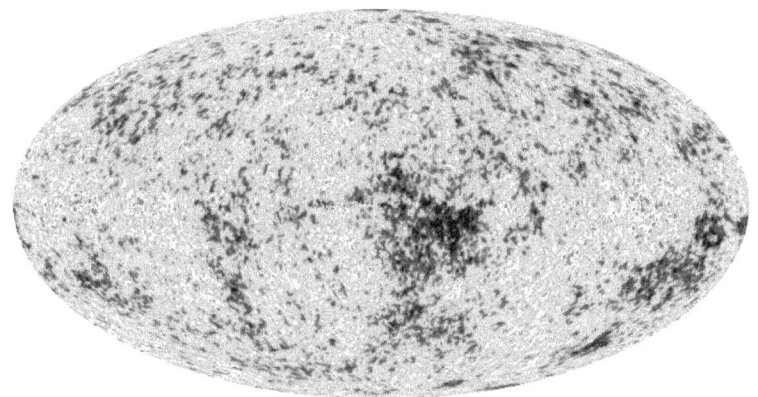

Fig.25: CMB (Cosmic Microwave Background)

> *"**The CMB is a snapshot of the oldest light in our Universe**, imprinted on the sky when the Universe was just 380,000 years old. It shows tiny temperature fluctuations that correspond to regions of slightly different densities, representing the seeds of all future structure: the stars and galaxies of today."*

CMB is a radiation coming from the farthest we can observe in Our Universe (the oldest light in Our Universe), because before it was dark. Then, **CMB is a radiation coming from the OBSERVABLE UNIVERSE BOUNDARY.** Although CMB is not a Unruh radiation, as it is the Black Hole "Hawking radiation".

CMB is the echo that comes from the beginning of the universe, that is, **the echo of the Big Bang.**

> *"**The Universe can be considered like a hollow enclosed space**, and the CMB is the heat that has been measured more accurately"* *("Before the Big-bang", Martin Bojowald, 2009).*

Following I relate a very short and relevant text **(Implicate Order, The Science Forum, April.2015)** that resume very clear the main aim of the present chapter:

"I am suspicious regarding that territory immediately beyond our **Hubble volume ["Observable Universe"]**, *but while it is natural to conclude that we can simply project further than this boundary,* **something tells me that part of the answer to our quest lies at or immediately beyond that boundary itself** *and it may just be forever beyond our empirical reach or alternatively considerably shake our current view on BB cosmology. But that is just a hunch of mine based on the principle that* **the boundary that exists between the quantum domain and the classical domain (the planck scale) might be translated macroscopically immediately at or beyond our classical macroscopic limit**. *The same* **principle applies at the event horizon of a black hole** *where a macroscopic object such as a black hole* **exhibits quantum properties at its boundary,** *and hence my interest in boundaries and also in exotic theories such as the* **holographic universe or the background independent Causal Dynamic Triangulation,** *which certainly would be a revolution to our thinking if any of these approaches gain momentum as a serious Quantum Gravity candidate. But following a resolution to the QG debate, what next, as a further onion skin will be peeled away."*

Some scientists have proposed that **Our Universe (3D space) could be a White Hole within a 4D space universe** *("bulk"). So our perception of it would be very limited.*

It would be like a prisoners inside a cave could only know what happens outside through the shadows produced by the light reflecting from a fire outside.

If ever they could leave the cave, their perception and understanding will greatly improve, although they also nor would understand everything.

HOLOGRAM THEORY

In a larger sense, the theory suggests that **the entire universe can be seen as a two-dimensional information on the cosmological horizon**, such that the three dimensions we observe are an effective description only at macroscopic scales and at low energies.

An important consequence is that the **maximum amount of information that can contain a region of space** surrounded by a differentiable surface **is limited by the total area of this surface**. It is similar what we have seen with Black-hole.

*Explained in an easier way, a **Holographic Theory of Our Universe** proposes that **everything that happens in Our Universe** (3D space) **could be explained** (or formulated) **by a border** (spatial 2D) Theory of Our Universe (**in the event horizon of Our Universe**).*

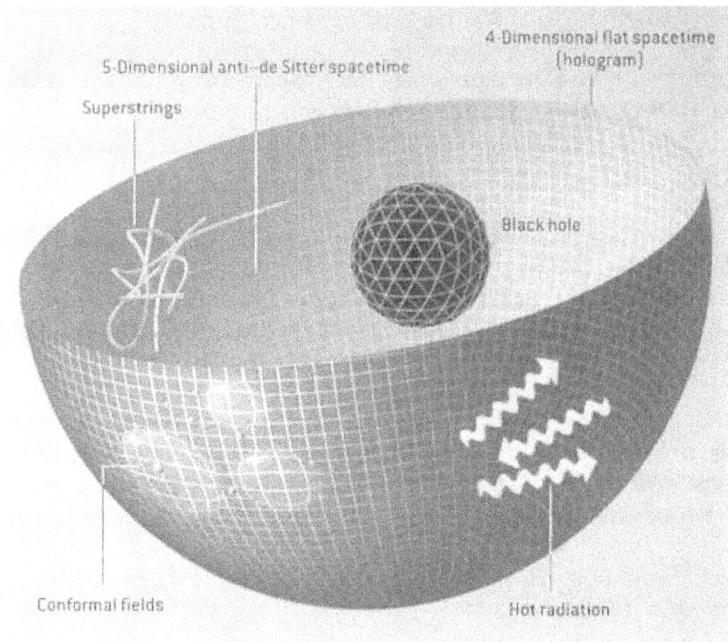

Fig.26: Hologram theory

Holographic Cosmology has not been mathematically precise because **the Cosmic Horizon of Our Universe is not well defined and grows over time (inflation).**

According to **John Maldacena** *(See article Scientific American, January-2006):*

"HOLOGRAM theory states that a quantum theory of gravity within a space-time anti-De Sitter is equivalent to a theory of ordinary particles at the border."

"Unfortunately not yet known any theory of boundary that results in an interior theory that includes just the four forces we observe in our universe [...] **Since our universe has not a defined boundary** *(such as having a space of anti-De Sitter and as precise holographic theory),* **we are not sure how a holographic theory for Our Universe would be defined due that there is no appropriate place to put the hologram.**"

One option could be to **propose, as a boundary of Our Universe for the HOLOGRAM theory,** that it will not be situated on higher scales (Cosmic Horizon), but it **could be on the smaller scales (Planck Horizon) where we could also have a 2D space boundary.**

Thus the border or limit of Our Universe would be on the Down (in the lower limits) of Our Universe, and not on the top which does not seem to have an established 2D space border. **Hologram principle would work DOWN to UP instead of UP to DOWN !**

Leonard Susskind, after reading the first Article (Oct.2012) tell me: *"...as you proceed down the scales to the smallest things, you place strings last. That's probably not right. The string scale if it exists must be bigger than the Planck scale, although not necessarily by a lot.* **At the Planck scale itself we should probably add the qubit. That's the unit of information at the horizon of a black hole.**"

A qubit is a quantum bit , the counterpart in quantum computing to the binary digit or bit of classical computing. Just as a bit is the basic unit of information in a classical computer, a qubit is **the basic unit of information in a quantum computer**.

Should we compare the information on surface of a black hole (measured in qubits) with the information existing on the lower scalar levels (Planck scale), and considering both as a 2D texture ?.

> **This 2D "virtual" surface at Planck scale could be the boundary to be considered for the HOLOGRAM theory: the Planck Horizon (Boundary).**

*The **US Fermi or Fermilab National Laboratory started (2014) a unique experiment that could reveal our three-dimensional world is just an illusion hologram.***

When we watch a movie on TV, we see images that seem integrity, real and three-dimensional. As we approach the screen we can see that the picture is actually a set of two-dimensional pixel.

Similarly believes astrophysicist particles Craig Hogan, who works at Fermilab, that everything we see with our eyes, the whole universe could be an illusion, composite images of 'pixels', ie, all information about our universe might be encoded in small packs of two dimensions. Would be so small that they can not be seen with the naked eye: about 10 billion billion times smaller than an atom.

*To this end Fermilab has constructed an instrument called **holometer**. This is a sensitive device that could allow these tiny discern 'pixels'.*

*To date, **the experiment is in the process of collecting information.***

*We know that **energy on the atomic level**, for instance, **is not continuous and comes in small, indivisible amounts**. The Holometer was built to test if space and time behave the same way. But results tell us what our common sense, and the laws of physics, assumes that **space and time are continuous.** (At least until the studied distances).*

*However, these studies are not yet definitive or conclusive (see **https://holometer.fnal.gov** and **arXiv.org/gr-qc/1512.01216**)*

9. FRACTALS & SCALE RELATIVITY

Physical models have evolved over the history (Aristotle, Newton, Einstein ...) covering every time a wider scale spectrum of the Universe.

Currently, various theories (Superstring, Quantum Gravity, ...) try to encompass all the scales of Our Universe (from Planck's scale to the limits of Our Universe), unifying both Relativity and Quantum theories (hitherto incompatible), but in addition, they pretend also to expand this scale spectrum out of these limits (above and below of Our Universe). They are known as **TOE**, and, as the name suggests, **they are trying to cover (parametrize) all the physics events of the Global Universe (Whole Universe).**

> "String theory is a wonderful theory that has already led to the discovery of deep mathematical and physical ideas, but it is very difficult to find the connection to the real world."
> "The problem of string theory is that it is defined on a scale of energies that it is ten thousand billion (10 e +13) times than that we can explore experimentally with current instruments."
>
> Lisa Randall (2005),"Wrapped passages"

Both, QM (all things are interconnected) and Einstein Relativity (all things are relational), seems to be telling us the same thing, albeit from different perspectives, which gives hope for a unification.

> *Fractal Theory could be another model that could help us to parametrize a wider spectrum of the Whole Universe.*

A **fractal** is a geometric figure that is **divided into smaller versions of _"it-self"._** Every fractal has an initiator and generator. The **initiator** represents the first step of the fractal and the **generator** produces each phase or step of the fractal. _(See Annex 3)_

To describe the Universe we have to be able to extrapolate and to create **"new smaller versions of _"other models"_ but linked by some kind of underla-ying patterns".**

To model Our **3D** Universe (and the **nD** Whole Universe), we have to consider/ use **3D-nD fractals.**

Fig.27: Escher´s circle as Fractal & Hologram sample

SCALE DIMENSION & FRACTAL DIMENSION:

If we consider the dimensions as the degrees of freedom of a system (e.g. the 4 dimensions of Our Universe: **Space Dimensions**: XYZ and **Temporal Dimension**: T). But **, sometimes, we also should consider the Scale Dimension: S** (as another "dimension") to locate an event within <u>Our Universe</u> (e.i.: one <u>molecule</u> could be in the **same Space-Time Dimension** than a <u>planet</u>, but if we don´t know the exact Scale Dimension we are considering, possibly, **we will not be able to see or find one of them**).

It is clear that this **Scale Dimension will be dependent of the other 3 space dimensions (X-Y-Z)**, and possibly we only must consider the "precision/accuracy" of the X-Y-Z values (e.g. it will be not the same to say X = A (in 10 e +10 units) than X = A (in 10 e -10 units). **Both can describe the same spacial location but at different scales (precision/accuracy)**: the first one the Earth scale, and second one the molecule scale.

<u>**Scale Dimension**</u> could be related with the **<u>Fractal Dimension,</u>** as *the index for characterizing fractal patterns or sets by quantifying them as a ratio of the change in detail by the change in scale.*

Such familiar scaling relationships can be defined mathematically by the general scaling rule in below Equation, where the variable N stands for the **number of new sticks**, ϵ **for the scaling factor**, and D **for the fractal dimension**:

$$N \propto \epsilon^{-D}$$

The symbol $\propto$ denotes proportionality.

FRACTAL COSMOLOGY

Fractal Cosmology is a set of minority **cosmological theories which state that the distribution of matter in the Universe, or the structure of the universe itself, is a fractal across a wide range of scales**. More generally, it relates to the usage or appearance of fractals in the study of the universe and matter. A central issue in this field is the fractal dimension of the universe or of matter distribution within it, when measured at very large or very small scales.

Demostration of Fractality for large scale of the universe requires additional observations (particularly of microwaves background radiation) and complicated mathematical solutions based on the theory of relativity of Einstein, which presents great complexity. Some of their most ambitious goals, **fractality of the universe could determine with unprecedented degree of accuracy, the distribution of galactic superclusters and generally all matter in the universe, including the dark.**

In **theoretical cosmology fractal geometry has been used as an attempt to describe the irregular nature should have spacetime at very small scale due to quantum fluctuaciones.** This has been conjectured that at very small scales spacetime is not smooth and has structure of differentiable manifold but should be a kind of **"quantum foam."**

In this context we have tried to explain the collapse of space-time that occurs inside **black holes** and relate proton gravity level, beating some of the biggest pitfalls of the current cosmology. **This model could make corrections to the model of Big Bang**.

Finally, mathematics conjetures raised about the alleged fractal nature of quantum mechanics, **getting the exotic idea of sacrificing time for a mono-dimensional and two-dimensional fractal time.**

A Fractal Universe? (Robert L. Oldershaw, 2002, http://www3.amherst.edu/~rloldershaw/NOF.HTM*)*

"ABSTRACT: From subatomic particles to superclusters of galaxies, nature has a nested hierarchical organization. There are also suggestive hints that s*elf-similarity, the idea of similar form on different size scales,* **might be a fundamental property of the cosmological hierarchy**. These features are the hallmarks of fractal structure. **Could nature, as a whole, be a fractal system?**

POSSIBLE IMPLICATIONS OF COSMOLOGICAL SELF-SIMILARITY:
If the dark matter is composed of ultra-compact stellar scale objects with a mass spectrum that is approximated by predictions of the self-similar hypothesis, then **it would appear that discrete self-similarity is a newly identified global property of nature. This would certainly change our current understanding of the cosmos.** Firstly it would provide a new approach toward a more unified understanding of nature, since **cosmological self-similarity implies analogous physics on all observable scales**. It would also imply that the usual assumption that **the universal hierarchy has cutoffs at about our current observational limits**, an assumption that has always

seemed suspiciously anthropocentric, should be questioned. **If cosmological self-similarity is verified, then it would seem more likely that additional scales underlie the atomic scale and encompass the galactic scale.** According to the new paradigm [...], **a new fractal geometry of space-time-matter would appear to be called for.**

If microlensing experiments verify the unique predictions mentioned above, however, we would still be faced with some important and very difficult questions. **How many scales are there in all, a finite number or "worlds within worlds" without end?** How strong is the degree of self-similarity between analogues? **Why is nature self-similar, and why are scales separated by a factor of about 5x10^{17}?** Like past discoveries, this one too would come wrapped in enigmas.

Some might argue that the self-similar cosmological paradigm is too fantastic to be true, that it is too speculative to deserve serious attention. But is it more fantastic or speculative than Alice In Wonderland theories like cosmic strings, shadow matter, the "many worlds" inter-pretation of quantum mechanics, etc. Probably not, if judged objectively, and at least the self-similar model can make definitive predictions and point to actual observational support. It is possible that nature really does involve the "worlds within worlds" structure of a fractal system. **Certainly there is enough supporting evidence to warrant serious consideration of discrete cosmological self-similarity. And soon, via microlensing experiments, we will learn nature's own verdict on this hypothesis.**"

SCALE RELATIVITY THEORY

Scale relativity theory is a **geometrical and fractal space-time theory.** The idea of a fractal space-time theory was first introduced by Garnet Ord, and by Laurent Nottale, in a paper with Jean Schneider. The proposal to combine fractal space-time theory with relativity principles was made by Laurent Nottale. The resulting scale relativity theory **is an extension of the concept of relativity found in special relativity and general relativity to physical scales** (time, length, energy, or momentum scales). In physics, relativity theories have shown that position, orientation, movement and acceleration cannot be defined in an absolute way, but only relative to a system of reference.

Noticing the relativity of scales, as noticing the other forms of relativity is just a first step. **Scale relativity theory proposes to make the next** step by translating this simple insight formally in physical theory, by **introducing explicitly in coordinate systems the "state of scale".**

To describe scale transformations requires the use of fractal geometries, which are typically concerned with scale changes. **Scale Rela-**

tivity is thus an extension of relativity theory to the concept of scale, using fractal geometries to study scale transformations.

The **principle of relativity** says that physical laws should be valid in all coordinate systems. This principle has been applied to states of position (the origin and orientation of axes), as well as to the states of movement of coordinate systems (speed, acceleration). **Scale relativity proposes,** in a similar way**, to define a scale relative to another one, and not in an absolute way.** Only scale ratios have a physical meaning, never an absolute scale, in the same way as there exists no absolute position or velocity, but only position or velocity rates.

If Einstein showed that space-time was curved, Nottale shows that it is not only curved, but also fractal. It means that such a space depends on scale. (See Annex 5)

Scale Relativity vs DSR:

Both theories have identified the Planck length as a fundamental minimum scale. However, as Nottale comments: *"the main difference between the "Doubly-Special-Relativity" approach and the Scale Relativity one is that **Scale Relativity have identified the question of defining an invariant length-scale as coming under a relativity of scales.**"*

Scale Relativity And Fractal Space-Time: A New Approach to Unifying Relativity and Quantum Mechanics. 2011 1st ed. World Scientific Publishing Company (Laurent Nottale, 2011).

*"This book provides a comprehensive survey of the development of the **theory of scale relativity and fractal space-time**. It suggests an original solution to the dis-unified nature of the classical-quantum transition in physical systems, enabling the basis of quantum mechanics on the principle of relativity, provided this principle is extended to scale transformations of the reference system. **In the framework of such a newly generalized relativity theory** (including position, orientation, motion and now scale transformations), **the fundamental laws of physics may be given a general form that unifies** and thus goes beyond the **classical and quantum regimes** taken separately. A related concern of this book*

is the **geometry of space-time, which is described as being fractal and non differentiable**. *It collects and organizes theoretical developments and applications in many fields, including physics, mathematics, astrophysics, cosmology and life sciences."*

Some other articles from the same author:

THE THEORY OF SCALE RELATIVITY (Laurent Nottale, 1991):

*"Basing our discussion on the **relative character of all scales in nature and on the explicit dependence of physical laws on scale in quantum physics, we apply the principle of relativity to scale transformations.** This principle, in combination with its breaking above the Einstein-de Broglie wavelength and time, leads to the demonstration of **the existence of a universal, absolute and impassable scale in nature, which is invariant under dilatation**. This lower limit to all lengths is identified with **the Planck scale, which now plays for scale the same role as is played by light velocity for motion**. We get new scale transformations of a Lorentzian form and generalize the de Broglie and Heisenberg relations. As a consequence the high energy, length and mass scales now decouple, **energy and momentum tending to infinity when resolution tends to the Planck scale**, which thus plays the role of the previous zero point.*
*This theory solves the problem of divergence of charge and mass (self-energy) in electrodynamics, **implies that the four fundamental couplings (including gravitation) converge at the Planck energy**, improves the agreement of GUT ("Grand Unification Theory") predictions with experimental results, **and allows one to get precise estimates of the values of the fundamental coupling constants."***

Scale relativity and fractal space-time: theory and applications (Laurent Nottale, 2009):

*"...during the last decades, the various sciences have been faced to an ever increasing number of new unsolved problems, of which many are linked to questions of scales. It therefore seemed natural, in order to deal with these problems at a fundamental and first principle level, **to extend theories of relativity by including the scale in the very definition of the coordinate system,** then to account for these scale transformations in a relativistic way."*

10. OUR UNIVERS CONSTANTS

Our Universe today has some fundamental constant values that govern its basic principles (*"The Constants of Nature"* ,John D. Barrow, 2002):

```
G =      Gravitational constant = 6,67.10 e -11  N.m2 / kg.s2
c =      Speed of light = 300,000 km/s
h =      Planck's constant = 6,63.10 e-34 J.s
e =      Electron charge = 1,6.10 e-19 C
Mpr =    Mass proton (neutron) = 1,67.10 e-27 kg
Npr =    Number of protons in Our Universe = 10 e +80
```

And the former constants lead to other fundamental constants or relationships:

```
α =   Fine structure constant = 2пe2 / hc = 1/137
αG =  Heavy structure constant = G.Mpr2 / hc = 10 exp -38
β =   Ratio between proton and electron mass = Mpr / I = 1,840
```

But it is not known if these values have always been constant, or if they have varied (evolved) since the Big Bang to our days, and if they will continue to evolve in the future.

Another "constant" value to consider is the **Cosmological Constant (Λ = 10 e-116 J)** . It is the energy that produces the current expansion of the universe (it could be equivalent to the vacuum energy). *We know that this value itself has been changing since the Big-bang*.

These constants are the characteristic of Our Universe (nowadays), and for the current time since the Big-bang. **We are quite sure that they can be different in other "pocket" universes of the Cosmic Landscape, and also we cannot be sure that they all have remained constant ever since the Big-bang.**

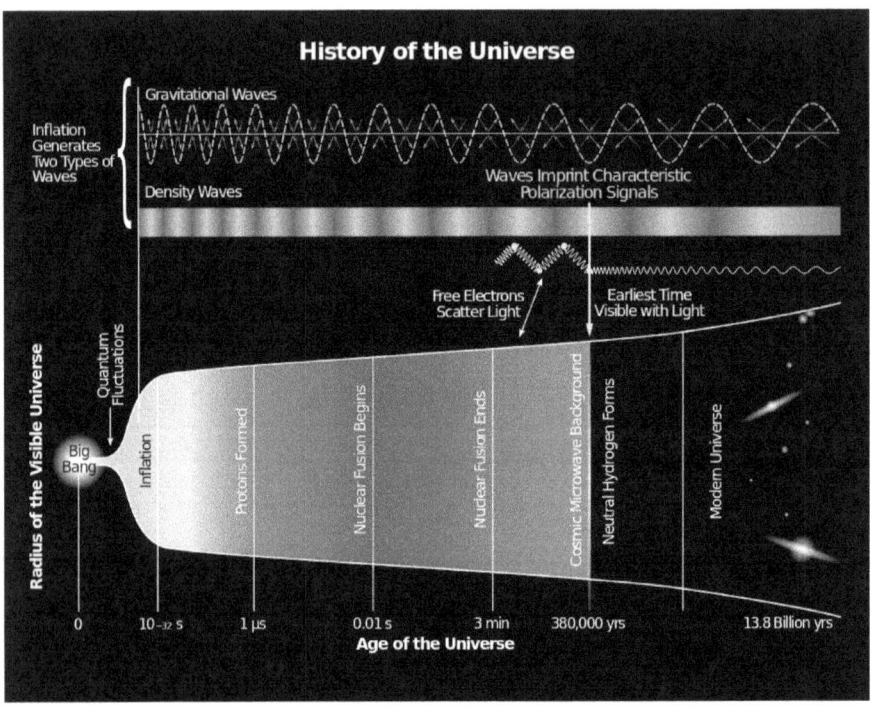

Fig.28: Cosmological Constant varied since the BB

Thus, so daily and usual constants for us as Gravitational (G) and speed of light (c) may be different in other "pocket" universes, and even they may have varied in Our Universe from the Big-bang, or **they may be different for different scale spectra of Whole Universe**. And this can also happen with the other constants such as the number of protons (+ neutrons) of our universe, etc.

> - **Our Universe constants (G, c, ...) could vary with Space Scale (?).**
>
> - **On others Scale Landscapes (Cosmic, Planck,...) could appear new constants that we cannot foresee by now.**

And something similar could happen with the other constants, such as the number of protons (+ neutrons) of Our Universe, the masses and electric charges of protons and electrons, etc.

Can we consider that the **number of protons and electrons have been the same since the Big Bang?** Will they be the same in the future?. Or, do both, protons and electrons, increase with the expansion of Our Universe? In the latter case, do they self-generate from nothing? , Or do they come from outside of Our Universe ?.

On the other hand, are there particles such as the electron and the proton in other Universes? Or **will be there other similar particles but with different masses and electrical charges?,** and forming a chemistry totally different from the one we know?

These **fundamental "constants"** serve to understand and parameterize Our Universe, and for the present moment of evolution (expansion), but **may be different for other Universes, or for other instants of evolution (expansion) of Our Universe.**

EDDINGTON´S UNCERTAINTY CONSTANT

Even they are, possibly, outside of the scope of this article, I would like to introduce in this section a reference to some**"uncertainty" values or constants of Our Universe.** Although they are not scientifically proven and accepted (and also they are not considered by mainstream), they have been considered, studied and defined by two great physical-mathematical and from different times (eras): **Pythagoras and Sir Arthur Eddington** (*"The Crystal Sun", Robert Temple, 2000*). **These values gives us an idea of a possible deviation between the theoretically expected and reality.**

> **_The Comma of Pythagoras_** (VI century BC), **_equivalent to 1.0136._** This constant reveals the difference in sound that occurs at the end of seven-eighths (= 7 x 6 tones = 42), with that sound what occurs at the end of twelve fifths (= 12 x 3,5 tones = 42), that should theoretically be exactly the same, but in practice it is not (so the theoretical value of the Coma should be exactly = 1).

Uncertainty Constant of Sir Arthur Eddington (proposed in his latest posthumously book **_"Fundamental Theory", 1953)_**, **equivalent to 9,604 x 10 e -14,**. **This constant measure also the discrepancy between reality and theoretically expected values** (The real values measured vs theoretical values calculated). Arthur Eddington argued that deviation by the fact that **the real physical coordinate frame has an standard deviation "sigma" over the pure and theoretical geometric frame.** Because, in the theoretical frame its origin is a pure geometric point, while in the real frame its origin has a probability distribution "sigma" from the theoretical point.

This explanation of Arthur Eddington, reminds us to the proposed QM & string theory, in which a **point/particle is replaced by a wave function ("sigma" function) or a string-brane**.

This constant might show us that , possibly, **our reality may not be based on a fixed and absolute physical coordinates.** On the contrary, in **Our Universe could be some uncertainties underlying on its own imprecise essence** (Uncertain Principle, Strings-branes, particle-wave essence, wave function,...).

11. GÖDEL THEOREME VS TOE

As we have seen, currently **there is an incompatibility between Relativity Theory** (which explains the physical phenomena of large scales) **and Quantum Theory** (which allows us to understand the physical phenomena of the smaller scales).

TOEs or Theories of Everything, like String Theory or M Theory, **are physical theories that try to unify both theories,** giving a global explanation to the physical phenomena of all the scales of the Universe.

Some scientists (eg, S. Hawking) believe that **Gödel's Incomplete Theorem implies that any attempt to construct a Theory of Everything is doomed to failure.**

"Up to now, most people have implicitly assumed that there is an ultimate theory, that we will eventually discover. Indeed, I myself have suggested we might find it quite soon. However, M-theory has made me wonder if this is true. **Maybe it is not possible to formulate the theory of the universe in a finite number of statements.** *This is very reminiscent of* **Gödel's theorem**. *This* **says that any finite system of axioms, is not sufficient to prove every result in mathematics."** *(Stephen Hawking: Gödel and the end of physics, 2002).*

http://www.damtp.cam.ac.uk/events/strings02/dirac/hawking/

Stephen Hawking was originally a believer in a Theory of Everything (TOE), but after accepting Gödel's theorem, he concluded that it could not be obtained. Until he realized the implication of Gödel's incompleteness theorem, he implicitly assumed that a TOE would be found, probably relying on what might be termed "scientific intuition." According to S. Hawking, the positivist philosophy of science is that every good physical theory is a mathematical model.

And since, **according to Gödel's Incomplete Theorem,** there are mathematical results that cannot be tested, then there must also be **physical theories that cannot be tested, including TOEs.**

Gödel's Incompleteness Theorem essentially says that generally (under certain conditions) **mathematical (physical) theories are inconsistent and incomplete.** And it can be concluded that any **Mathematical Theory (or Physics) can not be demonstrated by its own axioms.**

Gödel's Incompleteness Theorems **establish certain limitations on what can be demonstrated by mathematical reasoning.** *To speak precisely about what "can be demonstrated" or not, we study a mathematical model called* **formal theory** *consisting of a series of signs and rules forming the formulas, and certain successions of formulas that generate demonstrations. The theorems of a certain theory are then all formulas that can be demonstrated from a certain initial collection of formulas that are assumed as* **axioms**.

The first incompleteness theorem *states that, under certain hypotheses (which is an* **arithmetic theory** *- on natural numbers - and* **recursive teory** *- using algorithms), a* **formal theory cannot be both complete** *(answer any questions)* **and consistent** *(no contradictions present).*

The second incompleteness theorem *states that* **if the axiom system of the Theory in question is consistent, it is not possible to prove it by such axioms**. *This is a particular case of the former.*

In this book we have proposed that **TOEs could be just theories that attempt to cover a wide range of dimensional scales of the entire spectrum of the Universe** (eg, from 10 e-35 to 10 e + 30 m). But every time this spectrum expands, new laws and concepts will emerge (new Landscapes). So we would need other laws and we will require the development of new models and patterns to understand these new physical landscapes/spectra. This would imply a modification or extension of the previous TOEs. **If the Universe spectra was infinite, a single TOE would be impossible.**

Possibly the use of **Fractal Theory might be a system that allows to avoid this problem** (Gödel's incompleteness), to be able to use the **Fractal Theory as a "main" system of axioms**, which **simply set which model** (system of axioms) **we ought to use to each landscape** (scale range) as reference model (**based on some underlying patterns**).

Then we could have several systems of axions:

- Different models systems for different scale levels (Landscapes): **Landscape ToE System.**
- A **Fractal Model System** that could describe when to use the different ToEs (within the different scale levels).
- And, possibly, some **Underlaying Laws System** that could link the previous Systems (ToEs & Fractal).

It will be a multi-system of axions !

If we consider as a sample the ***Game of Life, devised by the British mathematician John Horton Conway in 1970:***

It is based on a very simple and basic principles. But as larger entities are formed, also begin to appear other concepts, such as movement, shooting, bombing, reproduction, ... And these concepts can be explained by other laws that have (only a very underlaying relation) to the initial law.

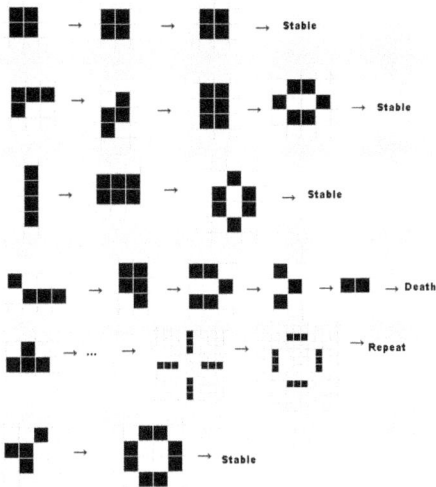

*But in this sample we start with some very elemental and basic initial objects (black and white squares) and rules (change black & white squares). These will be the underlaying laws of the game. **There will not be any object & laws more elemental below them.** But yes there could enlarge the objects and to get higher laws.*

Fig.29: Game of Life (John Horton Conway , 1970)

May we imagine that it can also happen in our universe? May we believe that there will also be some basic and elementary objects (and laws), since which there will be nothing smaller? **Only, if this were so, then we could think on getting a TOE based on these elementary and basic objects and laws.**

---

*It may happen that the **laws of nature have no boundaries** (they have an infinite range, and we can never know them in their totality), or that they are bounded (it is the opinion of RP Feymann). In the latter case, two things can happen: either we get to know all the Laws of Nature (TOE), or that the experiments are increasingly complex and expensive, and **we can only get to know 99.99% of The phenomena (Fractal Theory).***

PART III

This book part contain 5 annex with more detailed explanation about some of the (newest) theories we have dealt in the other parts:

ANNEX 1: MOND THEORY
ANNEX 2: EMERGENCE THEORY
ANNEX 3: FRACTAL THEORY
ANNEX 4: BRANES THEORY
ANNEX 5: SCALE RELATIVITY THEORY

Comparing and relating them with the main proposal of this book: **the nD-Fractal-Scale-Emergence Rainbow**

ANNEX 6: GR + QM + TOE THEORIES

Finally we include a **summary about the current mechanic theories (Classic, Relativistic and Quantic)**, as well the **TOE** (Theories of Everything).

ANNEX 1: MOND THEORY

In this annex it will be shown some theories that propose to modify the current Newton and Relativistic Theories for large scales (>10 e +20 m), that could explain some effects that appear there (the movement of galaxies,…).

It will be explained in short the MOND theory, and other similar theories, and how it could be linked to the aim of this book (the Fractal Scale Relativity of the Universe).

MOND BASIC PRINCIPLES

Modified Newtonian dynamics (**MOND**) is a theory that proposes a modification of Newton's laws to account for observed properties of galaxies. Created in 1983 by Israeli physicist Mordehai Milgrom, **the theory's original motivation was to explain the fact that the velocities of stars in galaxies were observed to be larger than expected based on Newtonian mechanics.** Milgrom noted that this discrepancy could be resolved if the gravitational force experienced by a star in the outer regions of a galaxy was proportional to the square of its centripetal acceleration (as opposed to the centripetal acceleration itself, as in Newton's Second Law), or alternatively **if gravitational force came to vary inversely with radius** (as opposed to the inverse square of the radius, as in Newton's Law of Gravity). **In MOND, violation of Newton's Laws occurs at extremely small acceleration**s, characteristic of galaxies yet far below anything typically encountered in the Solar System or on Earth.

MOND is an example of a class of theories known as modified gravity, and **is an alternative to the hypothesis that the dynamics of galaxies are determined by massive, invisible dark matter halos.** Since Milgrom's original proposal, MOND has successfully predicted a variety of galactic phenomena that are difficult to understand from a dark matter perspective. However, MOND and its generalizations do not adequately account for observed properties of galaxy clusters, and no satisfactory cosmological model has been constructed from the theory.

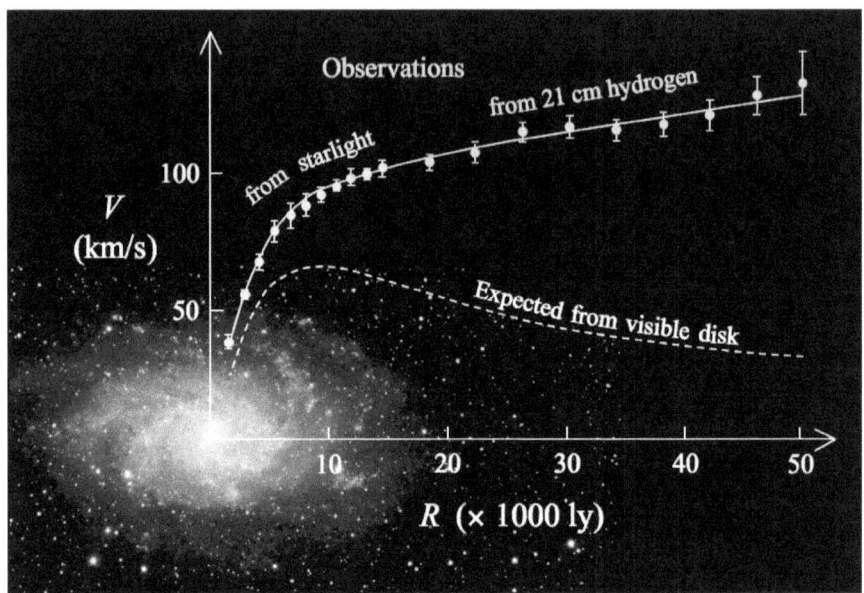

Fig.30: Comparison rotation curves (galaxy M33)

Comparison of the observed and expected rotation curves of the typical spiral galaxy M33

The basic premise of MOND is that while **Newton's laws** have been extensively tested in high-acceleration environments (in the Solar System and on Earth), they **have not been verified for objects with extremely low acceleration**, such as stars in the outer parts of galaxies. This led Milgrom to postulate a new effective gravitational force law (sometimes referred to as **"Milgrom's law"**) that relates the true acceleration of an object to the acceleration that would be predicted for it on the basis of Newtonian mechanics.[1] This law, the keystone of MOND, is chosen to **reduce to the Newtonian result at high acceleration but lead to different ("deep-MOND") behaviour at low acceleration**.

$$\mathbf{F_N} = m\mu\left(\frac{a}{a_0}\right)\mathbf{a}$$

Here **F_N** is the Newtonian force, m is the object's (gravitational) mass, **a** is its acceleration, μ(x) is an as-yet unspecified function (known as the "interpolating function"), and a_0 is a new fundamental constant which marks the transition between the Newtonian and deep-MOND regimes.

Agreement with Newtonian mechanics requires **μ(x) → 1 for x >> 1,** and consistency with astronomical observations requires **μ(x) → x for x << 1.** Beyond these limits, the interpolating function is not specified by the theory, although it is possible to weakly constrain it empirically.[

Milgrom's law can be interpreted in two different ways.

1) One possibility is **to treat it as a modification to the classical law of inertia** (Newton's second law), so that the force **F** on an object is not proportional to the particle's acceleration (F=m.**a**) but rather to **F=m μ(a/a₀)a**. In this case, the **modified dynamics would apply not only to gravitational phenomena, but also those generated by other forces, for example electromagnetism.**
2) Alternatively, Milgrom's law can be viewed as **leaving Newton's Second Law intact** and instead modifying the inverse-square law of gravity, so that the true gravitational force on an object of mass **m** due to another of mass **M** is roughly of the form (F=) **GMm/(μ(a/a₀)r²)**. In this interpretation, Milgrom's modification **would apply exclusively to gravitational phenomena.**

By itself, **Milgrom's law is not a complete and self-contained physical theory,** but rather an ad-hoc empirically-motivated variant of one of the several equations that constitute classical mechanics. Its status within a coherent non-relativistic theory of **MOND is akin to Kepler's Third Law within Newtonian mechanics**; it provides a succinct description of observational facts, but must itself be explained by more fundamental concepts situated within the underlying theory. Several complete classical theories have been proposed (typically along "modified gravity" as opposed to "modified inertia" lines), which generally yield Milgrom's law exactly in situations of high symmetry and otherwise deviate from it slightly.

A subset of these **non-relativistic theories have been further embedded within relativistic theories**, which are capable of making contact with non-classical phenomena (e.g.,gravitational lensing) and cosmology. Distinguishing both theoretically and observationally between these alternatives is a subject of current research.

The majority of astronomers, astrophysicists and cosmologists accept ΛCDM (based on General Relativity, and hence Newtonian mechanics), **and are committed to a dark matter solution** of the missing-mass problem. MOND, by contrast, is actively studied by only a handful of researchers.

The primary **difference between supporters of ΛCDM and MOND** is in the observations for which **they demand a robust, quantitative explanation and those for which they are satisfied with a qualitative account**, or are prepared to leave for future work. Proponents of MOND emphasize **predictions made on galaxy scales** (where MOND enjoys its most notable successes) and believe that a cosmological model consistent with galaxy dynamics has yet to be discovered; proponents of ΛCDM (Dark Matter) require high levels of cosmological accuracy (which concordance cosmology provides) and argue that a resolution of galaxy-scale issues will follow from a better understanding of the complicated baryonic Known matter) astrophysics underlying galaxy formation.

OUTSTANDING PROBLEMS FOR MOND

The most serious problem facing Milgrom's law is that **it cannot completely eliminate the need for dark matter** in all astrophysical systems: galaxy clusters show a residual mass discrepancy even when analysed using MOND. The fact that some form of unseen mass must exist in these systems detracts from the elegance of MOND as a solution to the missing mass problem, although **the amount of extra mass required is 5 times less than in a Newtonian analysis**, and there is no requirement that the missing mass be "non-baryonic" (Dark).

The 2006 observation of a pair of **colliding galaxy clusters known as the "Bullet Cluster"**,poses a significant challenge for all theories proposing a modified gravity solution to the missing mass problem, including MOND. Astronomers measured the distribution of stellar and gas mass in the clusters using visible and X-ray light, respectively, and in addition mapped the inferred dark matter density using gravitational lensing. **In MOND, one would expect the missing mass** (which is only apparent since it results from using incorrect Newtonian as opposed to MONDian dynamics) **to be centred on the visible mass. In ΛCDM**, on the other hand, **one would expect the dark matter to be significantly offset from the visible mass** because the halos of the two colliding clusters would pass through each other (assuming, as is conventional, that dark matter is collisionless), whilst the cluster gas would interact and end up at the centre. **An offset is clearly seen in the observations**. It has been suggested, however, that MOND-based models may be able to generate such an offset in strongly non-spherically-symmetric systems, such as the Bullet Cluster.

Several other studies have noted observational difficulties with MOND. For example, it has been claimed that MOND offers a poor fit to the velocity dispersion profile of globular clusters and the temperature profile of galaxy clusters, that <u>different values of a_0 are required for agreement with different galaxies' rotation curves,</u> and that **MOND is naturally unsuited to forming the basis of a theory of cosmology.**

Besides these observational issues, **MOND and its generalizations are plagued by theoretical difficulties.** Several ad-hoc and inelegant additions to general relativity are required to create a theory with a non-Newtonian non-relativistic limit. The plethora of different versions of the theory offer diverging predictions in simple physical situations and thus make it difficult to test the framework conclusively, and some formulations (most prominently those based on modified inertia) have long suffered from poor compatibility with cherished physical principles such as conservation laws.

TENSOR-VECTOR-SCALAR GRAVITY (TEVES)

Tensor–vector–scalar gravity (**TeVeS**), developed by Jacob Bekenstein in 2004, is a **relativistic generalization of Mordehai Milgrom's Modified Newtonian dynamics (MOND)** paradigm.

The main features of TeVeS can be summarized as follows:

- As it is derived from the action principle, **TeVeS respects conservation laws;**
- In the gravity weak-field approximation of the spherically symmetric, static solution, **TeVeS reproduces the MOND acceleration formula;**
- **TeVeS avoids the problems** of earlier attempts **to generalize MOND**, such as superluminal propagation;
- As it is a relativistic theory **TeVeS can accommodate gravitational lensing.**

MOND does not apply to cosmological scale for the same reasons that does not apply the theory of Newton. A covariate version, as Einstein's general relativity is required. **The most accepted version covariate is reduced to small-scale MOND theory called Theory Jacob Bekenstein Tensor-Vector-Scalar** (or TeVeS).

MODIFIED GRAVITY (MOG) THEORY

Ker Than, Oct.2007, SPACE.COM
(See Link: http://www.space.com/4554-scientists-dark-matter-exist.html)

Two Canadian astronomers think there is a good reason **dark matter,** a mysterious substance thought to make up the bulk of matter in the universe, **has never been directly detected: It doesn't exist.**

Dark matter was invoked to explain how galaxies stick together. **The visible matter alone in galaxies—stars, gas and dust—is nowhere near enough to hold them together**, so scientists reasoned there must be something invisible that exerts gravity and is central to all galaxies.

August 2007, an astronomer at the University of Arizona at Tucson and his colleagues reported that a **collision between two huge clusters of galaxies** 3 billion light-years away, known as the Bullet Cluster, **had caused clouds of dark matter to separate from normal matter**. Many scientists said the observations were proof of dark matter's existence and a serious blow for alternative explanations aiming to do away with dark matter with modified theories of gravity.

But **John Moffat**, an astronomer at the University of Waterloo in Canada, and **Joel Brownstein**, his graduate student, **say those announcements were premature.**

In a study detailed in the Nov. 2007 issue of the Monthly Notices of the Royal Astronomical Society, the pair says their **Modified Gravity (MOG) theory** can explain the Bullet Cluster observation. MOG differs from other modified gravity theories in its details, but is similar in that **it predict that the force of gravity changes with distance.**

"MOG gravity is stronger if you go out from the center of the galaxy than it is in Newtonian gravity," Moffat explained. "The stronger gravity mimics what dark matter does. With dark matter, you take Einstein and Newtonian gravity and you shovel in more dark matter. If there's more matter, you get more gravity. Whereas for me, **I say dark matter doesn't exist. It's the gravity that's changed.**"

Using images of the Bullet Cluster made by the Hubble, Chandra X-ray and Spitzer space telescopes and the Magellan telescope in Chile, **the scientists analyzed the way the cluster's gravity bent light from a background galaxy**—an effect known as gravity lensing. The pair concluded that **dark matter was not necessary to explain the results.**

*"Using **Modified Gravity theory**, the 'normal' matter in the Bullet Cluster is enough to account for the observed gravitational lensing effect,"* Brownstein said. *"Continuing the search for and then analyzing other merging clusters of galaxies will help us decide **whether dark matter or MOG theory offers the best explanation for the large scale structure of the universe.**"*

Moffat compares the modern interest with dark matter to the insistence by scientists in the early 20th century on the existence of a **"luminiferous ether",** a hypothetical substance thought to fill the universe and through which light waves were thought to propagate. *"They saw a glimpse of special relativity, but they weren't willing to give up the ether,"* Moffat told SPACE.com. *"Then Einstein came along and said we don't need the ether. The rest was history."*

Douglas Clowe, the lead astronomer of the team that linked the Bullet Cluster observations with dark matter (and now at Ohio University), says he still stands by his original claim. For him and many other astronomers, **conjuring up new particles that might account for dark matter is more palatable than turning a fundamental theory of how the universe works on its head:**

"As far as we're concerned, [Moffat] hasn't done anything that makes us retract our earlier statement that the Bullet Cluster shows us that we have to have dark matter," Clowe said. ***"We're still open to modifying gravity to reduce the amount of dark matter, but we're pretty sure that you have to have most of the mass of the universe still in some form of dark matter."***

Fig.31: Image of the Bullet Cluster

A composite image of the Bullet Cluster, a much-studied pair of galaxy clusters that have collided head on. One has passed through the other, like a bullet traveling through an apple, and is thought to show clear signs of dark matter (blue) separated from hot gases (pink).

Credit: X-ray: NASA/ CXC/ CfA/ M.Markevitch, Optical and lensing map: NASA/ STScI, Magellan/ U.Arizona/ D.Clowe, Lensing map: ESO/WFI

FRANCISCO R. VILLATORO. NAUKAS BLOG, 2011 *(http://francis.naukas.com/2011/03/02/por-que-la-teoria-mond-requiere-la-existencia-de-la-materia-oscura/)*

*"Mordehai Milgrom, proposed in 1983 that **Newton's laws were not correct on a galactic scale**, called Modified Newtonian Dynamics (MOND), which helps to explain the galactic rotation curves*

using only the visible matter (without using the dark matter that Fritz Zwicky introduced to explain the orbital velocities of galaxies in galaxy clusters and Vera Rubin curves applied to the speed of stars in galaxies). **MOND "works well for the 'small' scale of individual galaxies, but does not tell you much about the universe at large scales of galaxy clusters and above**, where the universe is well described by the theory of dark matter".

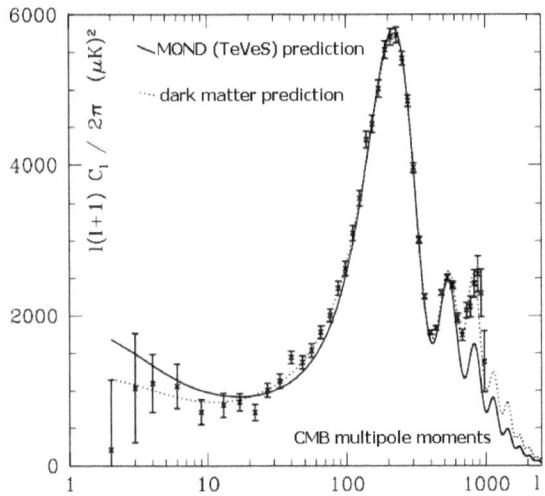

Fig.32: MOND (TeVeS) & DarkMatter predictions.

"But **MOND does not imply that dark matter does not exist. The MOND theory does not explain the anisotropy in the cosmic microwave background (CMB)** that are sensitive to the existence of dark matter in the early universe. Multipolar odd spectrum peaks are higher WBC and the couple peaks are lower in early universe dominated by dark matter, as opposed to a market dominated by ordinary matter (as in the original theory MOND) universe. Actually the "theory" **MOND is not a theory applicable to <u>cosmological</u> level**; it has to use the theory **TeVeS Bekenstein** (the most accepted covariant version of the "theory" MOND). As the figure shows, MOND (TeVeS) contradicts the data obtained by the WMAP satellite to the CMB, so MOND (without dark matter) is an incorrect theory (at least cosmological scale). Therefore, many MOND specialists physicists believe that **dark matter exists and is an integral part of MOND, but it is not necessary to explain the <u>galactic rotation curves</u>**; dark matter in MOND **is only necessary to the <u>cosmological</u> scale** to explain the dynamics of galaxy clusters and superclusters and CMB."

MOND THEORIES VS FRACTAL RAINBOW

The **different physic theories** that propose **changes** to the current theories **for the different scale landscapes of the universe** (for large scales: MOND TeVeS, MOG,...; and for small scales: DSR,...) , may be **theories to be considered as potential emerging theories or laws that vary according to the scale.**

MOND modifies Newton's law to very low accelerations, so we should be able **to connect this acceleration with some distance** (possible **distance to the center of the galaxy**).

We should see if the laws of Newton-Einstein suffer variations (changes) depending on the spatial scale of reference:

• **Newton's (& Einstein) law** seems to work perfectly to explain the dynamic phenomena at **scales up to the Solar System** (**max. 10 e +15 m**),.

• From this distance we must consider the **MOND´s law**, which should fully explain the dynamic phenomena at scales between that **10 e+15 to 10 e +20 m (galaxies)** , specifically, the rotation of the stars from a distance to the center of the galaxy.

• Should be **above these scale (10 +20 m)**, to the **ends of Our Universe (10 +27 m)**, where the **other dynamic laws (TeVeS´ law)** could describe better these scales phenomena.

• And **above these scales (> 10 e +30 m)** other **new emergent laws** will apear to explain new emergent phenomena that we cannot forecast by now.

As we already state in chapter 5:

The **Dynamic Laws of Physics (and Universal Gravitation)** have varied over time, and even Einstein had already proposed that they still has to evolve:

ARISTOTLE: $F = m.v$
NEWTON: $F = m.a$
EINSTEIN. $E = m.c^2$ (*)
MOND: $F = m.a.(A/A_0)$
FRACTAL RAINBOW: F = f (scale) = m.a.(scale factor)

(*) This equation does not correspond to the same dynamic concept but has many similarities.

Other option to be considered is that **G (Gravitational Constant) could also vary with the SCALE:**

> Then the attraction (**F**) between two masses (**M** and **m**) separate by distance (**d**) will be:
>
> $$F= G.M.m/d^2$$
>
> Where **G = f (Scale) = f (d)**, but only to be relevant for large scales or distances.

This variation of G could be due to the **Fractality of Spacetime**:

"*Applying relativity to fractal non-differentiable spacetime, Laurent Nottale, in his* **Scale Relativity theory**, *suggests that potential energy arises due to the fractality of space, and accounts for the missing mass-energy observed at cosmological scales*".

This missing of mass-energy could make also that G (gravity constant) decrease at cosmological scales. Gravity could be missed through the fractal space-time.

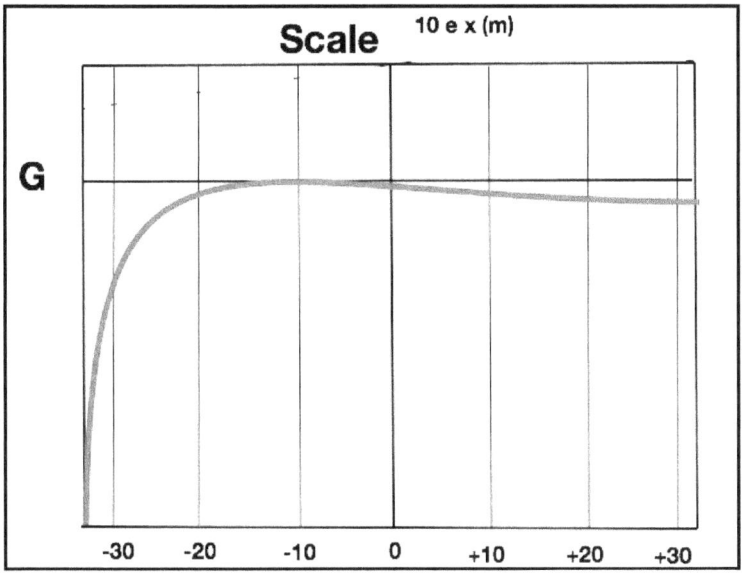

Fig.33: Variation of G (Gravity Constant) with Scale

ANNEX 2: EMERGENCE THEORY

In this annex it will be explained in short the Emergence theory, and other similar theories, and how it could be linked to the aim of this book (the Fractal Scale Relativity of the Universe).

Emergence is one of the main concepts within the aim of the proposal of this book.

EMERGENCE CONCEPT

The central message of the book **"A different universe" (R. Laughlin, 2007)** is **that real frontier of science is not in the small, but in the complex.** When many atoms are added to form a solid or a biological tissue, there are new organizational principles are not rigorously derived from microscopic laws and meaningless in low particulate systems.

Laughlin insists that the investigation of **these concepts that operate in the complex organization of matter is so "fundamental" research as the elemental forces**. If you go to science as a whole, the idea of emergency, under which the whole is more than its parts, is much more relevant than the reduction, since almost all scientific activity, including physical, they are dealing with emerging technology concepts. These range from temperature of a liquid, to the resistance of a building, through a flower morphology. In the "different universe" proposed by Laughlin, **science is reconciled with common sense because all our perception of reality is based on <u>emerging concepts and laws.</u>**

Emergence is a process whereby larger entities, patterns, and regularities arise through interactions among smaller or simpler entities that themselves do not exhibit such properties.

135

STRONG AND WEAK EMERGENCE

Usage of the notion "emergence" may generally be subdivided into two perspectives, that of **"weak emergence"** and **"strong emergence"**. In terms of physical systems, **weak emergence is a type of emergence in which the emergent property is amenable to computer simulation**. This is opposed to the older notion of **strong emergence**, in which **the emergent property cannot be simulated by a computer**.

Some common points between the two notions are that **emergence concerns new properties produced as the system grows**, which is to say ones which are not shared with its components or prior states. Also, it is assumed that the properties are supervenient rather than metaphysically primitive (Bedau 1997).

Weak emergence describes new properties arising in systems as a result of the interactions at an elemental level. However, it is stipulated that t**he properties can be determined by observing or simulating the system**, and not by any process of a priori analysis.

We speak of **weak emergency** when there are properties that are identified as emerging by an external observer but **may be explained by the properties of the primary constituents of the system**. In the case of crystallization water molecules: crystal qualities belong neither to hydrogen or oxygen, but they can be explained and predicted from them

Fig.34: Example of emergence in a physical system:

The formation of complex symmetrical and fractal patterns by Snowflakes is an example of weak emergence in a physical system.

Strong emergence describes the direct causal action of a high-level system upon its components; qualities produced this way are irreducible to the system's constituent parts (Laughlin 2005). **The whole is other than the sum of its parts.** An example from physics of such emergence is water, being seemingly unpredictable even after an exhaustive study of the properties of its constituent atoms of hydrogen and oxygen. It follows then that no simulation of the system can exist, for such a simulation would itself constitute a reduction of the system to its constituent parts (Bedau 1997).

However, *"the debate about whether or not the whole can be predicted from the properties of the parts misses the point.* **Whole produce unique combined effects, but many of these effects may**

137

be co-determined by the context and the interactions bet-
ween the whole and its environment(s)" (Corning 2002).

"The ability to reduce everything to simple fundamental laws
does not imply the ability to start from those laws and re-
construct the universe. *The constructionist hypothesis breaks*
down when confronted with the twin <u>*difficulties of scale and comple-*</u>
<u>*xity*</u>*.* *At each level of complexity entirely new properties ap-*
pear. *Psychology is not applied biology, nor is biology applied che-*
mistry. We can now see that **the whole becomes not merely**
more, but very different from the sum of its parts." *(Anderson*
1972).

The plausibility of strong emergence is questioned by some
as contravening our usual understanding of physics.

Mark A. Bedau observes: *"Although strong emergence is logically*
possible, it is uncomfortably like magic. How does an irreducible but
supervenient downward causal power arise, since by definition it
cannot be due to the aggregation of the micro-level potentialities?
Such causal powers would be quite unlike anything within our scien-
tific ken. This not only indicates how **they will discomfort reaso-**
nable forms of materialism. *Their mysteriousness will only heigh-*
ten the traditional worry that emergence entails illegitimately get-
ting something from nothing."

Meanwhile, **others have worked towards developing analytical**
evidence of strong emergence. In 2009, ***Gu et al.*** presented a
class of **physical systems that exhibits non-computable ma-**
croscopic properties. They concluded that

"Although macroscopic concepts are essential for understanding our
world, much of fundamental physics has been devoted to the search
for a `theory of everything', a set of equations that perfectly descri-
be the behavior of all fundamental particles. The view that this is
the goal of science rests in part on the rationale that such a theory
would allow us to derive the behavior of all macroscopic concepts, at
least in principle. The evidence we have presented suggests that
this view may be overly optimistic. ***A `theory of everything' is***
one of many components necessary for complete understan-
ding of the universe, but is not necessarily the only one. *The*
development of macroscopic laws from first principles may involve
more than just systematic logic, and ***could require conjectures***
suggested by experiments, simulations or insight"

Emergent structures are patterns that emerge via collective actions of many individual entities. To explain such patterns, one might conclude, per Aristotle, <u>**that emergent structures are other than the sum of their parts**</u> on the assumption that the emergent order will not arise if the various parts simply interact independently of one another. However, <u>there are those who disagree</u>. According to this argument, **the interaction of each part with its immediate surroundings causes a complex chain of processes that can lead to order in some form**. In fact, some systems in nature are observed to exhibit emergence based upon the interactions of autonomous parts, and some others exhibit emergence that at least at present cannot be reduced in this way.

EMERGENT PHYSIC SAMPLES

Some examples include:

- **Classical mechanics:** The laws of classical mechanics can be said to emerge as a limiting case from the rules of quantum mechanics applied to large enough masses. This is particularly strange since quantum mechanics is generally thought of as *more* complicated than classical mechanics.

> Obviously typical formulas of classical mechanics (dynamic: $F = m.a$), can not be extrapolated from the formulations of the much more complex quantum mechanics.

- **Friction:** Forces between elementary particles are conservative. However, friction emerges when considering more complex structures of matter, whose surfaces can convert mechanical energy into heat energy when rubbed against each other. Similar considerations apply to other emergent concepts in continuum mechanics such as viscosity, elasticity, tensile strength, etc.

> Friction, viscosity, elasticity, etc., they are considered "non-conservative" forces emerging from the different complexity associated matter (molecules).

- **Temperature** is sometimes used as an example of an emergent macroscopic behavior. In classical dynamics, a *snapshot* of the instantaneous momenta of a large number of particles at equilibrium is sufficient to find the average kinetic energy per degree of freedom which is proportional to the temperature. For a small number of particles the instantaneous momenta at a given time are not statistically sufficient to determine the temperature of the system. However, using the ergodic hypothesis, the temperature can still be obtained to arbitrary precision by further averaging the momenta over a long enough time.

Temperature is no more than a measure of the kinetic energy of the particles (molecules) contained in a space. The temperature concept does not make sense within a single molecule. Temperature simply emerges to be many molecules in the same space.

In some theories of particle physics, even such **basic structures as mass, space, and time are viewed as emergent phenomena**, arising from more fundamental concepts such as the Higgs boson or strings. In some interpretations of quantum mechanics, **the perception of a deterministic reality, in which all objects have a definite position, momentum**, and so forth, **is actually an emergent phenomenon**, with the true state of matter being described instead by a wavefunction which need not have a single position or momentum. **Chemistry** can in turn be viewed as an emergent property of the laws of physics. **Biology** (including biological evolution) can be viewed as an emergent property of the laws of chemistry. Similarly, **psychology** could be understood as an emergent property of neurobiological laws. Finally, free-market theories understand **economy** as an emergent feature of psychology.

EMERGENCE SUMMARY

From the perspective of **REDUCTIONISM**, the laws of physics are what drive the universe and come from nowhere and everything is derived from them. While from the perspective of **"EMERGENTISM"**, they are rules of collective behavior that arise from more primitive rules of conduct, and they are valid only for a bounded number of circumstances (Scale Landscape?).

- **Newton's Laws are** not fundamental but **emerging**, as they are the result of the aggregation of quantum matter, that makes, fluids and solids, macroscopic collective (or organization) phenomenon.

- The molecules, atoms and subatomic particles respond to the **laws of Quantum Mechanics,** which **are surely emerging laws** as a result of these particles do not behave like physical particles, if not rather like waves.
- **The field of quantum mechanical consist on wave of "nothing",** and tries to explain by means of **wave-particle duality,** a concept that does not exist, and that only serves to **explain unusual phenomena by means of words and concepts known**.
- The same applies **Uncertainty Principle**. The Newtonian notion that the position and velocity characterize an object is incorrect (although our scale to be so), and must be replaced by another concept called **wave function**.
- **But, what is a wave?** It is understood as a difference of potential of a value corresponding to a substance (or field). But this definition does not serve for the EM waves, due to there are not detected any substance that allows electromagnetic waves: "ether". In the same way we accept that is non-existent the medium by which waves of quantum mechanics move.
- Moreover, **The Emergence of conventional physical reality from quantum mechanics is difficult to understand**.
- **The gravity theory of Einstein (General Relativity) is not a Basic Law, but an Emerging Principle:** It is a collective property of matter, that constitutes the space-time, and whose accuracy is greater with increasing distances, and it is lower (or even zero) when they are short (or too short).
- The empty space (**vacuum**) has a similar spectroscopic quantum structure than ordinary solids and fluids. Studies in large particle accelerators (LHC) have helped us to understand that the empty space (vacuum) **is more like ordinary glass, than the ideal vacuum of Newton.**

EMERGENCE VS FRACTAL RAINBOW

The concept of emergence (laws and concepts) is on in the **very essence of the proposal FRACTAL RAINBOW.**

It is evident that **for each scale level (landscape) there are different concepts** (strings, particles, atoms, molecules, cells, planets, galaxies ...), and that we need to use **different physics laws/models t**o interpret and predict their behavior (QM, Chemistry, Biology, Newton, Thermodynamics, SR+GR, MOND,...).

It is also evident that among these laws (models) there are certain correlations, and that **these laws are not independent from each other**, but are related by certain underlying laws, which often can anticipate the other, but not always (due to **chaos theory**).

CHAOS THEORY

Chaos theory is the field of study in mathematics that studies the behavior of dynamical systems that are highly sensitive to initial conditions—a response popularly referred to as the **butterfly effect. Small differences in initial conditions** (such as those due to rounding errors in numerical computation) **yield widely diverging outcomes for such dynamical systems, rendering long-term prediction impossible in general.** This happens even though these systems are deterministic, meaning that their future behavior is fully determined by their initial conditions, with no random elements involved. In other words, **the deterministic nature of these systems does not make them predictable.** This behavior is known as **deterministic chaos**, or simply **chaos**.

The theory was summarized by Edward Lorenz as: Chaos: "When the **present determines the future, but the approximate present does not approximately determine the future.**"

Chaotic behavior exists in many natural systems, such as weather and climate, and **Chaos Theory has applications in several disciplines**, including meteorology, sociology, physics, computer science, engineering, economics, biology, and philosophy.

Normally the laws of a higher scale level (higher spatial scale) can be explained (interpreted) by the laws of the lower scale level (lower spatial scale). It is what might be called **bottom-up: from low to high complexity**. We can understand atoms and molecule laws (chemistry) by knowing its particle composition laws (electron, neutron and proton).

COMPLEXITY VS SIMPLICITY

There is no absolute definition of what complexity means; the only consensus among researchers is that there is no agreement about the specific definition of complexity. However, a characterization of what is complex is possible. **Complexity is generally used to characterize something with many parts where those parts interact with each other in multiple ways.** The study of these complex linkages at various scales is the main goal of **complex systems theory.**

In science, there are as of 2010 a number of approaches to characterizing complexity. Neil Johnson states that "even among scientists, there is no unique definition of complexity - and the scientific notion has traditionally been conveyed
using particular examples..." Ultimately he adopts the definition of **'complexity
science'** as **"the study of the phenomena which_emerge_ from a collection of interacting objects."**

Warren Weaver posited in 1948 **two forms of complexit**y: disorganized complexity, and organized complexity. Phenomena of **'disorganized complexity'**
are treated using **probability theory and statistical mechanics**, while **'organized complexity'** deals with phenomena that escape such approaches and
confront **"dealing simultaneously with a sizable number of factors which
are interrelated into an organic whole"**.

But also it could happen in reverse, that is, **top-down: from high
to low complexity**. We can forecast the particle composition laws
(electron, neutron and proton) by studying the behavior of the different atoms and molecules (that is what did **Dmitri Mendeléyev** to
develop the Periodic table of the elements).

We use normally second way (bottom-up) to understand the lower
laws and concepts, and usually **we forecast particles and behaviors of these low scales without being able to see them**. We
only can see by electron microscopes scales of 10 e-10 m. Although
we can detect lower scales (10 e-17 m) by the particle colliders to
high energy (LHC-CERN).

For higher scales, we have been able to see direct by telescope till
approx 10 e+20 m (nearest galaxies), but by astro-photography
(HUBBLE) till the farthest galaxies of Our Universe (10 e+ 26 m). So
**we are able to understand better the behaviors of the objects
and phenomena at these high scales**.

Most times we have been able to forecast concepts (Higgs boson,...)
and behaviors (Black Holes,...). But sometimes we are not able to
forecast what will happen in other (higher or lower) scales landscapes, because its behavior is so complex to calculate (the weather,...)
or because appear new behaviors we didn´t forecast (due to chaos
principle).

> *I would like the reader think what will happen if humans live in one electron of hydrogen (with only one electron and one proton)instead of the Earth:*
>
> - **Electron size** *is 10 e-18 m* **and proton (nucleus) size** *will be 10 e-15 m* **(1.000 times larger)**.
> - *While* **Earth size** *is 10 e+7 m and* **Sun size** *is 10 e+9 (* **100 times larger**).
> - **The electron orbits around the nucleus** *with an orbit of 10 e-11 m* **(10.000 times the size of nucleus)** .
> - *While* **Earth orbit around the Sun** *is 10 e+10 (***10 times the size of the Sun**), *and* **Plutón 10 e +12 m (1.000 times).**
>
> *Both systems are so similar (!), but* **they are separates 10 e+25 scale times (order of magnitude).**
>
> **If we live on an electron, Will we see or forecast stars and galaxies ?**
>
> *If nowadays we are able tu understand till 10 e+27 m from the Earth (10 e +20 times the Earth size), proportionally, from the electron we will be able to understand 10 e+20 times electron size:* **Only 10 e+2 m (100 meters !).**

CONCLUSIONS

From the previous considerations, **we can conclude the following main concepts:**

- **Even scale spectrum (landscape), could be dominated by laws or events that,** although they may be explained by the underlying laws of lower spectra, **they cannot always be extrapolated ("a priori") from them.**
- **We could say that concepts** (such as energy, matter, space, time, speed,...), **and theories or laws** (such as Newton and Maxwell theory, thermodynamics, ...) are only valid for the human scale spectrum, but they **may have no meaning, as such, in other higher or lower scalar spectra.**
- **To quantum scales** (10 e -20 to 10 e -35 m) these concepts could be meaningless, and possibly **(1st and 2nd) Laws of Thermodynamics could not be longer valid.** They are Emergent Laws and Concepts.
- Possibly **at higher scale levels** (> 10 and +30 m), these laws and concepts also no longer have felt, and **appear other laws and concepts that better explain the events that occur.**
- **Also the time** is believed to be a consequence of heat (entropy), and he **only makes sense in those scales in which heat can be considered, and in which the laws of thermodynamics can be applied.**

- **<u>Gravity</u> can also be considered as an effect of entropy,** characteristic of our scale. **It is an emerging and not fundamental force.** *Gravity emerges when matter and mass emerge.*
- **The <u>vacuum</u> could be considered as another type of space phase.** Although, for the point of view of our scale, in a vacuum there is "nothing", that is not true for quantum scales, where **quantum fluctuations** are detected, that **can be considered as emerging effects typical of these scales.**
- The nature **laws could be endless (infinite scope) or they could be bounded** (it is the opinion of RP Feymann). In the latter case, two things could happen: either that **we get to know all the Laws of Nature (TOE),** or that experiments are increasingly complex and expensive, and **we only get to know the 99.99% of the phenomena (Fractal Theory).**

ANNEX 3: FRACTAL THEORY

In this annex it will be explained in short the Fractal theory, and how it could be linked to the aim of this book (the **Scale Relativity of the Fractal Universe**).

Fractal Theory is one of the main concepts within the aim of the proposal of this book. It is supposed to be a **mathematical tool that could help to parametrize the whole Universe scales.**

FRACTAL CONCEPT

A **fractal** is a natural phenomenon or a mathematical set that exhibits a repeating pattern that displays at every scale. It is also known as **expanding symmetry** or **evolving symmetry**. If the replication is exactly the same at every scale, it is called a **self-similar pattern**. Fractals can also be **"nearly" the same** at different levels. Fractals also include the idea of a detailed **pattern that repeats itself**

According to Falconer, rather than being strictly defined, **fractals should, in addition to being nowhere differentiable.**

Self-similarity, which may be manifested as:

- **Exact self-similarity:** identical at all scales; e.g. Koch snowflake:

Fig.35: Exact self-similarity

- **Quasi self-similarity:** *approximates the same pattern at different scales; may contain small copies of the entire fractal in distorted and degenerate forms; e.g., the Mandelbrot set's* **satellites are approximations of the entire set, but not exact copies.**

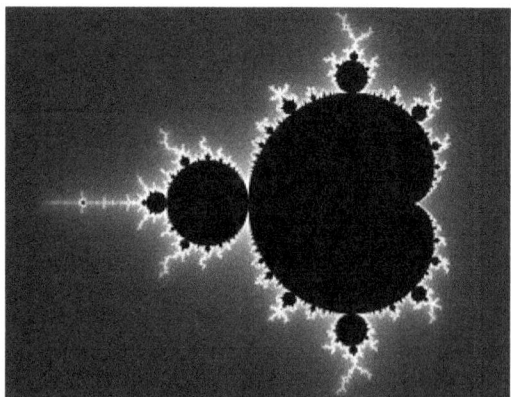

Fig.36: Quasi self-similarity

- **Statistical self-similarity:** repeats a pattern stochastically so numerical or statistical measures are preserved across scales; It is the weakest type of self-similarity: it demands that the fractal has numerical or statistical measures which are preserved with the change in scale. Random fractals are examples of fractals of this type.[·

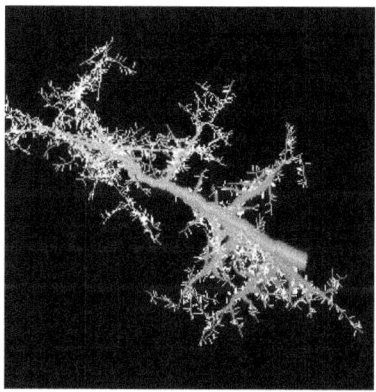

Fig.37: Statical self-similarity

- **Multifractal scaling:** characterized by more than one fractal dimension or scaling rule

Fig.38: Multifractal scaling

Fractals can have emergent properties, when fine or detailed structures are generated in arbitrarily small scales, generated by randomly or complex patterns. Oddly, in principle that the Fractal

Theory can only generate deterministic forms and processes, this is not so, and depending on the rules (repetition and similarity) used, **they can generate totally unexpected and different processes and forms, to the initial ones.**

"Long Is the Coast of Britain? Statistical Self-Similarity and Fractional Dimension" is a paper by mathematician Benoît Mandelbrot, first published in *Science* in 1967. In this paper Mandelbrot discusses self-similar curves that have Hausdorff dimension between 1 and 2. These curves are examples of *fractals*, although Mandelbrot does not use this term in the paper, as he did not coin it until 1975. The paper is one of Mandelbrot's first publications on the topic of fractals.

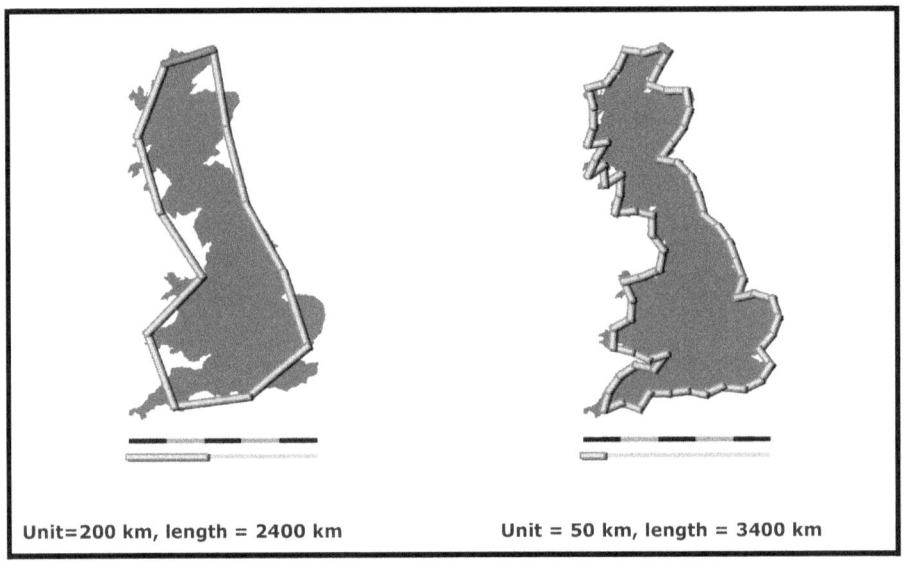

Unit=200 km, length = 2400 km   Unit = 50 km, length = 3400 km

Fig.39: Measuring the Coast of Britain

This paper is important because it is a "turning point" in Mandelbrot's early thinking on fractals. It is an example of the linking of mathematical objects with natural forms that was a theme of much of his later work. **In it is shown as the actual measurement of the Brittany cost increases with decreasing accuracy (precision) measurement.**

FRACTAL COSMOLOGY

In physical cosmology, **fractal cosmology** is **a set of minority cosmological theories which state that the distribution of matter in the Universe, or the structure of the universe itself, is a fractal across a wide range of scales** (multifractal system). More generally, it relates to the usage or appearance of fractals in the study of the universe and matter. A central issue in this field is the fractal dimension of the universe or of matter distribution within it, when measured at very large or very small scales.

The first attempt to **model the distribution of galaxies with a fractal pattern** was made by Luciano Pietronero and his team in 1987, and a more detailed view of the universe's large-scale structure emerged over the following decade, as the number of cataloged galaxies grew larger. Pietronero argues that the universe shows a definite fractal aspect over a fairly wide range of scales, **with a fractal dimension of about 2**. The fractal **dimension of a homogeneous 3D object would be 3**, and 2 for a homogeneous surface, whilst the **fractal dimension for a fractal surface is between 2 and 3**. The ultimate significance of this result is not immediately apparent, but it seems to indicate that **both randomness and hierarchal structuring are at work on the scale of galaxy clusters and larger.**

In the realm of theory, the first appearance of fractals in cosmology was likely with Andrei Linde's **"Eternally Existing Self-Reproducing Chaotic Inflationary Universe"** theory (Chaotic inflation theory), in 1986. In this theory, the evolution of a scalar field creates peaks that become nucleation points which cause inflating patches of space to develop into "bubble universes," **making the universe fractal on the very largest scales**. Alan Guth's 2007 paper on **"Eternal Inflation and its implications"** shows that this variety of Inflationary universe theory is still being seriously considered today. And inflation, in some form or other, is widely considered to be our best available cosmological model.

Since 1986, however, quite a large number of different cosmological theories exhibiting fractal properties have been proposed. And while Linde's theory shows fractality at scales likely larger than the observable universe, theories like **Causal dynamical triangulation and Quantum Einstein gravity are fractal at the opposite extreme, in the realm of the ultra-small near the Planck scale.** These recent theories of quantum gravity describe a fractal structure for spacetime itself, and suggest that the dimensionality of space evolves with time. Specifically; **they suggest that reality is 2D at the Planck scale, and that spacetime gradually becomes 4D at larger scales.** French astronomer Laurent Nottale first suggested **the fractal nature of spacetime** in a paper on Scale Relativity published in 1992, and published a book on the subject of Fractal Space-Time in 1993.

French mathematician Alain Connes has been working for a number of years to reconcile Relativity with Quantum Mechanics, and thereby to unify the laws of Physics, using Noncommutative geometry. **Fractality also arises in this approach to Quantum Gravity.** An article by Alexander Hellemans in the August 2006 issue of Scientific American quotes Connes as saying that the next important step toward this goal is to **"try to understand how space with fractional dimensions couples with gravitation."** The work of Connes with physicist Carlo Rovelli suggests that **time is an emergent property or arises naturally,** in this formulation, whereas in Causal dynamical triangulation, choosing those configurations where adjacent building blocks share the same direction in time is an essential part of the 'recipe.' **Both approaches suggest that the fabric of space itself is fractal, however.**

FRACTAL THEORY & FRACTAL RAINBOW

At the moment that we propose a universe of infinite spatial scales, it seems that the **fractal theory is a good candidate to establish a model that could parameterize these different scales, which may include different sizes** (space, time,...), **different own emerging constants** and different **physical laws**, although, possibly, interconnected in some underlying way.

As we have seen, the vast majority of the concepts of nature and the cosmos, remain complex fractal parameters (snowflakes, crystals, sea coasts, galaxies, ..). And it also seems that the vacuum can have a fractal composition.

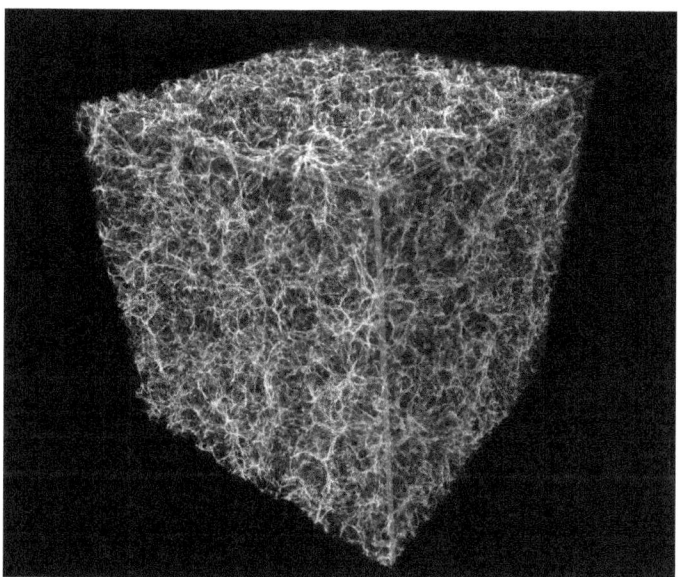

Fig.40: Matter distribution in a cubic section of the Universe.

(The blue fiber structures represent the matter (primarily dark matter) and the empty regions in between represent the cosmic voids).

We can compare Fig.38 and Fig.40, and we could see that both figures are very similar.

Also the latest physical theories seem based on fractals developments (CDT, Scale Relativity, ...).

This may be one of the research paths for future in physics and mathematics, which, today, and almost from 20 years ago, are too focused on string theory.

If something can anticipate it is that the fractal universe laws and patterns should be **complex, multi-fractals and multi-scalar.**

---

*"Why the universe is not a (simple) **fractal** (but **yes a multi-fractal**)".*
(Vincent J. Martinez & Bernard J.T. Jones,1990):

*"There is overwhelming evidence from the CfA redshift survey that the **distribution of galaxies in the Universe obeys a scaling law on length scales***

---

less than 5 Mpc/h. *Despite this scale invariance, the Universe is not well represented by a homogeneous fractal on these scales. The dependence of the correlation length r0 with sample depth and luminosity is studied. A method based on the minimal spanning tree is presented for determining the Hausdorff dimension, DH, of a point distribution. The technique is applied in order to find the Hausdorff dimension of the CfA redshift survey. The obtained value is DH = 2.1 +/- 0.1. The correlation dimension differs from this value, D2 = 1.3 +/- 0.1,* **therefore** the Universe **is not well characterized by only one exponent: it** is not a simple fractal. It is a more complex structure, a multi-fractal.***"*

OTHER RELATED ARTICLES OVER FRACTALS:

AN INFINITE FRACTAL COSMOS (*http://arxiv.org/pdf/ 1001.2865v1.pdf* - Robert L. Oldershaw. 2010)

"Centuries ago **Immanuel Kant, J. H. Lambert, Spinoza and a few others proposed an infinite hierarchical model of the universe based largely on natural philosophy.** *This general hierarchical paradigm never garnered a large following, but like the legendary Phoenix it kept arising from the ashes of neglect.* **In the 1800s and 1900s quite a few scientists, including E. E. Fournier d'Albe, F. Selety, C. V. L. Charlier and G. de Vaucouleurs, argued for hierarchical cosmological models based on the hierarchical organization within the observable universe.** *Then towards the end of the* **1970s,** *the mathematician* **B. B. Mandelbrot (1977, 1983) gave the hierarchical paradigm new life and widespread exposure by developing the mathematics of fractal geometry and demonstrating that fractal phenomena based on hierarchical self-similarity are ubiquitous in nature.** *In this way natural philosophers, research scientists, mathematicians and now theoretical physicists have all found their way, slowly but surely, to the infinite fractal paradigm.* **There are many routes to this paradigm, and certainly there are a large number of distinct versions (Oldershaw, 2001; Nottale, 1993; Tegmark, 2003; Baryshev and Teerikoorpi, 2003; Baryshev, 2008) of the basic paradigm, each with its own unique theoretical explanations for why and how nature is organized in this manner.** *For example, the present author has shown how a* further generalization of General Relativity involving discrete self-similar scaling of the interactions between matter and the geometry of the spacetime manifold leads to an unbounded discrete self-similar cosmos *(Oldershaw, 2007).*

Although we do not yet know how all of the technical details of the fractal cosmos will be resolved, **the general paradigm that nature is an infinite hierarchy of worlds within worlds has finally arrived, and it is likely to be our dominant paradigm for the foreseeable future."**

Summary from the book **"Infinite Universe in a Mote".** *(1994* **Written by Yun Pyo Jung)**

Gottfried Wilhelm Leibniz(1646 - 1716) *has suggested a unique idea called* **"Monadology".** *He thought that the universe consists of innumerable monads and another complete universe is concealed in each of them.*

*To consider this idea, you might start by understanding that **it represents a kind of fractal structure of the universe;** when a particle contains another complete universe in it, such a universe must be again composed of much smaller innumerable particles, in each of which another smaller universe may repeat.* **In a fractal structure, this process continues endlessly.**

If the universe were really formed in a fractal structure, you could say **our cosmos might be a particle, too. We may be living in a particle.** *Such particles as the cosmos may exist innumerably. And there may be a gigantic universe, and it may not be the end of all there is. In fact, it might be another particle in another greater universe. Such a process would also continue endlessly in a fractal structure.*

Stages Of The Universe

Let's call the large world inside the gigantic being 'the macro-world' and call the small world replicated in your body 'the micro-world'.

Then you can arrange all the stages of the universe from subatomic particles to the gigantic being as follows;

(1) Micro-world: *subatomic particles - (atomic nucleus) - atoms - molecules - macromolecules - morphological elements - cells - man*
(2) Macro-world: *stars(the sun) - (galaxial nucleus) - galaxies - clusters - great clusters - superclusters - the cosmos - gigantic being*

If the universe were replicated **in fractal structure, the magnitude ratios of corresponding elements between the two extreme worlds would be all constant.**

Atoms correspond to galaxies. *The idea that atomic structure may be similar to that of the solar system is not worth consideration at all.*

Constant Number 10^{30}

You have now calculated all ratios of corresponding elements between the micro-world and the macro-world. And **you may note all those are showing similar results containing a constant number of 10^{30}.**

If all these were not just a series of fortuity, **the universe could be said to be consecutive vertically in fractal structure**, and **the magnifying power between two adjacent levels of fractal to be around 10^{30}.**

Epilogue

If scientists accept this new theory and sort out what particular celestial bodies correspond to what particular particles in the micro-world, **exact comparisons will be possible without any variation.**

If scientists apply Fractal Cosmology to their research, **they may be able to get answers for most questions**, or obtain proper interpretations of various phenomena of the universe, by comparing the micro-world with the macro-world.

In the infinite universe, replications of fractal structure have neither a beginning nor an ending. But there exist levels in the fractal universe, and, **we understand now the magnifying power at each level is 10^{30}.**

Human science has progressed to the limits of their vision. **Fractal Cosmology may invite mankind into the world of infinity.**

REFERENCES TO KNOW MORE ABOUT FRACTALS:

FRACTAL EXPLORER (WEB):

http://www.wahl.org/fe/HTML_version/link/FE1W/c1.htm

ANNEX 4: BRANE THEORY

It is essential to include an annex on **Branes (String) Theory** which has led in recent years one of the most important scientific research fields in the search for a **Theory of Everything (TOE).**

Also in the proposal of this book many references and concepts are extracted from this theory: World-Brana, Multi-dimensional ...

BRANE CONCEPT

In string theory and related theories such as supergravity theories, a **brane** is a physical object that **generalizes the notion of a point particle to higher dimensions**. For example, a point particle can be viewed as a brane of dimension zero, while a string can be viewed as a brane of dimension one. It is also possible to consider higher-dimensional branes. In dimension p, these are called p-branes. The word brane comes from the word "membrane" which refers to a two-dimensional brane.

Branes are dynamical objects which can propagate through spacetime according to the rules of quantum mechanics. They have mass and can have other attributes such as charge. A p-brane sweeps out a $(p+1)$-dimensional volume in spacetime called its *worldvolume*. **Physicists often study fields analogous to the electromagnetic field which live on the world volume of a brane.**

In string theory, **D-branes are an important class of branes that arise when one considers open strings**. As an open string propagates through spacetime, its endpoints are required to lie on a D-brane. The letter "D" in **D-brane refers to a certain mathematical condition on the system known as the Dirichlet boundary condition.** The study of D-branes in string theory has led to important results such as the AdS/CFT correspondence, which has shed light on many problems in quantum field theory.

STRING THEORY (M-THEORY)

In physics, **string theory** is a theoretical framework in which the **point-like particles of particle physics are replaced by one-dimensional objects called strings.** String theory describes how these strings propagate through space and interact with each other. On distance scales larger than the string scale, a string looks just like an ordinary particle, with its mass, charge, and other properties determined by the vibrational state of the string. In string theory, one of the many vibrational states of the string corresponds to the graviton, a quantum mechanical particle that carries gravitational force. **Thus string theory is a theory of quantum gravity.**

String theory is a broad and varied subject that attempts to address a number of deep questions of fundamental physics. String theory has been applied to a variety of problems in black hole physics, early universe cosmology, nuclear physics, and condensed matter physics, and it has stimulated a number of major developments in pure mathematics. Because string theory potentially provides a unified description of gravity and particle physics, **it is a candidate for a theory of everything,** a self-contained mathematical model that describes all fundamental forces and forms of matter.

String theory was first studied in the late 1960s as a theory of the strong nuclear force, before being abandoned in favor of quantum chromodynamics. Subsequently, it was realized that the very properties that made string theory unsuitable as a theory of nuclear physics made it a promising candidate for a quantum theory of gravity. The earliest version of string theory, bosonic string theory, incorporated only the class of particles known as bosons. It later developed into **superstring theory**, which posits a connection called supersymmetry between bosons and the class of particles called fermions. Five consistent versions of superstring theory were developed before it was conjectured in the mid-1990s that they were all different limiting cases of a single theory in eleven dimensions known as **M-theory**. In late 1997, theorists discovered an important relationship called the **AdS/CFT correspondence**, which relates string theory to another type of physical theory called a quantum field theory.

One of the challenges of string theory is that the **full theory does not yet have a satisfactory definition in all circumstances.** Another issue is that the theory is thought to describe an enormous landscape of possible universes, and this has complicated efforts to develop theories of particle physics based on string theory. These issues have led some in the community to criticize these approaches to physics and question the value of continued research on string theory unification.

During the last 25 years, most scientists (and the corresponding research budgets) has focused on String Theory (and Branes), obtaining great advances and even very interesting proposals concerning the Universe (different universes and multiple dimensions, supersymmetry between particles and forces, world-branes at different scales, etc. But so many efforts seem to have reached to a "dead end". It has not been possible to establish a Theory of Everything clear and concrete, and simply it has been obtained different possible options that must be specified and tested. The next experiments at CERN can help us in this.

BRANE COSMOLOGY

Brane cosmology refers to several theories in particle physics and cosmology related to string theory, superstring theory and M-theory.

Brane and bulk:

The central idea is that **the visible, four-dimensional (3D+1T) universe is restricted to a brane inside a higher-dimensional space, called the "bulk" (also known as "hyperspace").** In the bulk model, at least some of the extra dimensions are extensive (possibly infinite), and other branes may be moving through this bulk. Interactions with the bulk, and possibly with other branes, can influence our brane and thus introduce effects not seen in more standard cosmological models.

One of the features that propose string theory (within the branes), is that by representing bosons causing different interactions (gravity, electromagnetism and nuclear weak and strong) as strings, it is proposed that the first **Gravity it could be a closed string** (so that it could travel between different branes, of different dimensions), while **the other three (EM, S and W) would open strings** and would be anchored to a membrane, unable to leave it.

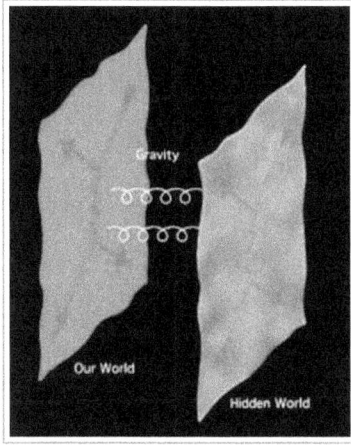

Fig.41: 2D Branes showing Gravity

Kaluza-Klein Theory (KK)

KK theory is a unified field theory of gravitation and electromagnetism built around the idea of a fifth dimension beyond the usual four of space and time. **It is considered to be an important precursor to string theory.**

The five-dimensional theory was developed in three steps:

- The original hypothesis came from Kaluza, who sent his results to Einstein in 1919, and published them in 1921. **Kaluza's theory was a purely classical extension of general relativity to five dimensions.** The 5-dimensional GR metric has **15 components**. Ten components are identified with the **4-dimensional spacetime metric**, 4 components with the **electromagnetic vector potential**, and one component with an unidentified scalar field sometimes called the **"radion"** or the **"dilaton"**. Correspondingly, **the 5-dimensional Einstein equations yield the 4-dimensional Einstein field equations, the Maxwell equations for the electromagnetic field, and an equation for the scalar field.**
- 1926, Oskar Klein gave Kaluza's classical 5-dimensional theory a quantum interpretation, to accord with the then-recent discoveries of Heisenberg and Schrödinger. **Klein introduced the hypothesis that the fifth dimension was curled up and microscopic.**
- It wasn't until the 1940s that the **classical theory was completed**, and **the full field equations including the scalar field were obtained.**

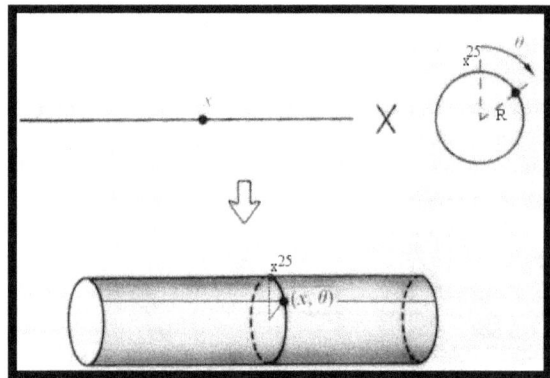

Fig.42: KK branes
(2D rolled surface on 1D X dimension)

A string 1D spatial, would have two other dimensions rolled, if we consider a rolled 2D surface hoselike. It is supposed that this scale will co-exist 6 dimensions (3 x 2D = 6D) plus the 3 we know (X-Y-Z). These 6D forms are called 6D Calabi-Yau shapes.

Calabi-Yau Branes

A **Calabi–Yau brane (shapes or manifold)**, also known as a **Calabi–Yau space**, is a special type of manifold that is described in certain branches of mathematics such as algebraic geometry. In superstring theory, **the extra dimensions of spacetime are sometimes conjectured to take the form of a 6-dimensional Calabi–Yau manifold**, which led to the idea of mirror symmetry.

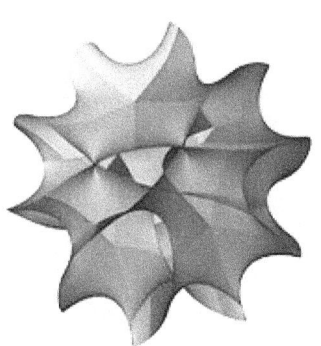 

Fig.43: Calabi-Yau Branes
Fig.44: Calabi-Yau Structure

There may be a lot of different 6D Calabi-Yau shapes. And all of them would form **the basic structure of space of Our Universe** (Our World-brane). The **"Quantum Foam"**.

> In the **CERN LHC particle collider,** specific experiments are **planned to detect these possible KK branes with extra dimensions, but to date they have not been detected.** With the maximum energies expected to be used (about 10 TeV) it is believed that we could detect these KK branes as long as they had dimensions greater than 10 e -20 m (100 times smaller than the size of the electrons).

Our 4D-Braneworld

We could suppose that **Our Universe** (Our 4D-Brane-world: 3D space and 1D Temporal) **may be floating within the nD-Bulk**, which would have more spatial dimensions (eg: 5D = 4D spatial and one Temporal); which, in turn, the bulk would contain other Universes Brane (Brane-world) of less dimensions.

And, moreover, in the smallest scales of Our Universe (Our Brane-world), they could have a compacted branes (6D Calabi-Yau) that form the basis of space. The **"empty" space would consist of different varieties of Calabi-Yau 6D, forming what we know as "quantum foam"**.

> *So the whole universe would consist on different D-Branes of different dimensions and size, floating or co-existing ones with the others. And where the different forces (interactions) and objects (energy-matter) may manifest differently on each one. Being required different models (laws) to parameterize them.*

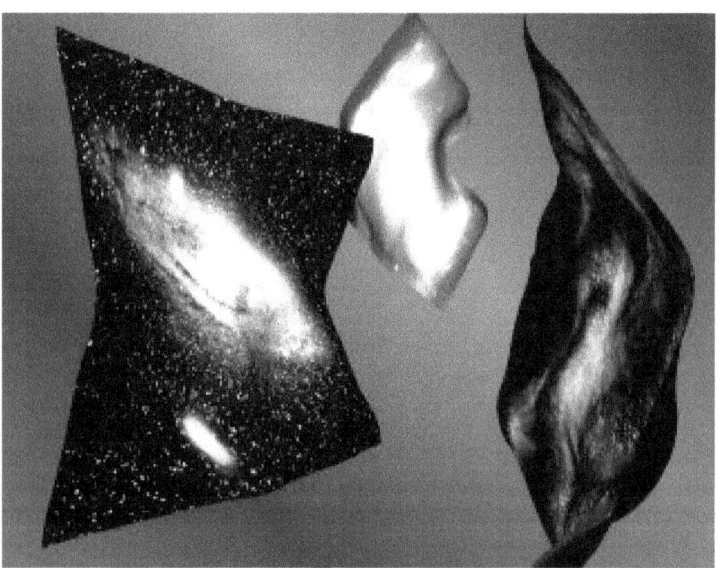

Fig.45: Multi D-Braneworlds in the Bulk

BRANE THEORY IN FRACTAL RAINBOW

The possibility that in the universe there may be other dimensions (and consequently the possibility of other parallel universes) is one of the implications of **String Theory**, which, to avoid inconsistencies (zero and infinite) that occur in the theory , this **requires the existence of more dimensions (usually 11 dimensions are proposed)**

Do not forget that, as proposed in this book (Fractal Rainbow), although **String Theory** can be a TOE proposal, it would **only cover a band of the scale spectrum of the Universe**: from the Planck scale (10 e -35 m), up to the limits of our universe, but also could cover the Cosmic Landscape (10 e + 35 m).

Then an option to describe this scale spectrum could be the proposal in the **section 2** of this book (***THE LIMITS OF OUR UNIVERSE. THE KA-LUZA-KLEIN THEORY: THE EXISTENCE OF NEW SPATIAL DIMEN-SIONS***). and shown in ***Figures 9 & 10, the Landscapes of Our Universe:***

- **Our Universe will be a 4D brane** (3D space+1 time) **floating on the "Bulk" (5D-brane)**, where could be other **"pocked universes"** similar to ours, but with different dimensions, constants and development status (explosion or implosion).
- Our Universe could contain **rolled 6D KK dimensions in a very small scales**, that are not able to be realized by us (this option could be demostrated during next experiments on LHC-CERN).

- *Here we will have the **11 dimensions: Our 4D Universe, + extra 1D "Bulk" + 6D rolled KK**.*
- *Other option is that the **"Bulk" could be a 11D** (or more), where **9D (space) have evolved to Our Universe**: 3D are large ones, and the other 6D are small and rolled.*
- ***Other "pocked" universes** could have evolved to different dimensions (e.i. 8D), and also they could have different large (e.i. 4D) or small (e.i. 4D) ones.*

Other thing to consider are the **INTERACTIONS FIELDS**.

If we accept the **Open and Close string theory** (explained in the previous section):
- **Only Gravity "closed" strings (force field and waves) could travel (affect) through the different D-Branes, between different Universes.** See attached figure where you can see the Newton Law applied for different dimensional spaces. Different Universes could be affected by the Gravity depending on their masses.
- **Electromagnetic "open" strings (EM field and waves) will be only trapped to one brane,**and they shall not to escape from it to

other branes (Bulk, KK,...).Then we will not be able to detect any EM wave from other universes from inside of Our own Universe.

- **Nuclear (Strong & Weak) "copen" strings (S+W fields and waves) will also be trapped to one brane.** Nuclear waves have still not been ever detected. But if they exist, sure they will be of very short scope, and only could be detected to a very small scales.

Assuming that the various world-branes (U1, U2, ...), with different constants (G, c, ...) and dimensions (D1, D2, ...), co-exsisted "floating" in the "Bulk" (Cosmic Landscape) that has N-dimensions and Gravitional Constant (Gpc), we can assume that between them will be an attractive Gravitional force proportional to its mass (and energy),and inversely proportional to the distance exponential N-1 (if N = 3 then e = 2 as in Our 3D Universe)

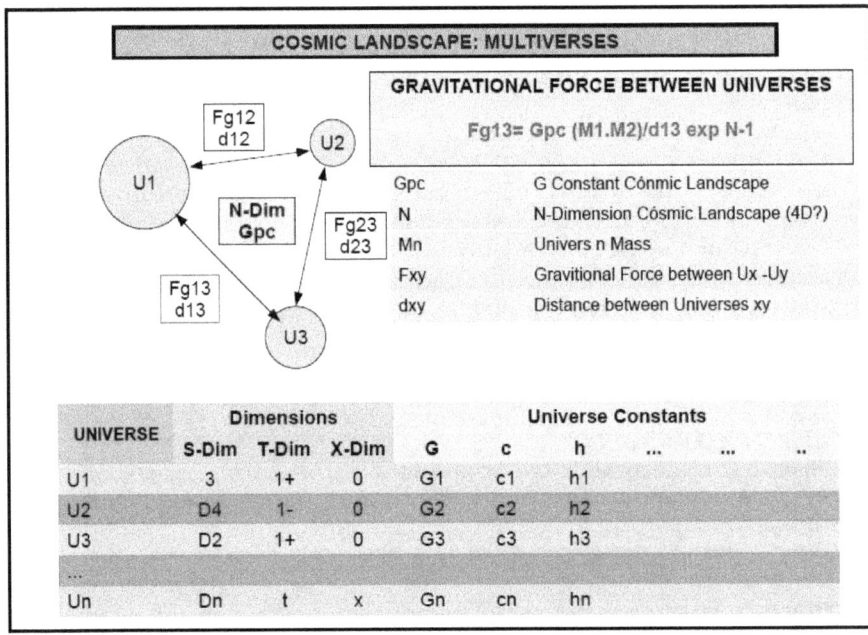

Fig.46: Gravity Interactions between D-Braneworlds

"May exist other worlds that we do not know in other branes separate from ours by other hidden dimensions".
"There could be an additional "filler" distributed between the different branes, or even the bulk, that could explain dark energy and dark matter".
"If there is life on another brane, these beings trapped in a completely different environment, should feel completely different forces (waves) that would be detected by different senses".
Lisa Randall (2005),"Wrapped passages"

ANNEX 5: SCALE RELATIVITY THEORY

In this annex it will be explained in short the **Scale Relativity** theory, and how it could be linked to the aim of this book (the **Fractal Rainbow**).

Scale relativity Theory consider a Fractal Space-Time and the need of consider the scale concept (Scale Factor) in physics laws, so, this theory seems to fit perfectly with the proposal of the Fractal Rainbow, but also we´ll see that there are some differences.

Theory of Scale Relativity, if someday is finally accepted by the "scientific community", it **could completely revolutionize the cosmological physics**. Instead, after 30 years of studies, developments and proposals (articles and books), Nottale, and his theory, are practically forgotten by the "scientific community", and his work marginalized of the academic domains (I knew of their existence in April 2015).

BASIC PRINCIPLES

Scale relativity is a geometrical and fractal space-time theory. The idea of a fractal space-time theory was first introduced by **Garnet Ord (*1983, "Fractal space-time: a geometric analogue of relativistic quantum mechanics". Journal of Physics A: Mathematical and General)*** , and by Laurent Nottale in a paper with Jean Schneider **(*1984, "Fractals and Nonstandard Analysis"* (PDF). *Journal of mathematical physics).*** The proposal to combine fractal space-time theory with relativity principles was made by **Laurent Nottale (*1989, "Fractals and the quantum theory of spacetime"*)**.

The resulting scale relativity theory is an extension of the concept of relativity found in special relativity and general relativity to physical scales (time, length, energy, or momentum scales). In physics, relativity theories have shown that position, orientation, movement and acceleration cannot be defined in an absolute way, but only relative to a system of reference.

Noticing the relativity of scales, as noticing the other forms of relativity is just a first step. Scale relativity theory proposes to make the next step by translating this simple insight formally in physical theory, by introducing explicitly in coordinate systems the "state of scale".

To describe scale transformations requires the use of fractal geometries, which are typically concerned with scale changes. Scale relativity is thus an extension of relativity theory to the concept of scale, using fractal geometries to study scale transformations. The construction of the theory is similar to previous relativity theories, with three different levels: Galilean, special and general.

Scale Relativity extends special and general relativity with a new formulation of scale invariance preserving a reference length, which postulates be the Planck length.

By requiring this length is invariant under changes of state of scale, it is necessary to abandon the hypothesis of differentiability of spacetime. Instead a spacial-time fractal structure is suggested.

The development of a full general scale relativity is not finished yet. However, the existing progress and results already have consequences for the foundations of quantum mechanics, particle physics, and high energy physics. Furthermore, empirical predictions in physics, astrophysics, and cosmology have already been validated, most often with a high precision, or highly statistically significant results.

THE PRINCIPLE OF RELATIVITY

The **principle of relativity** says that **physical laws should be valid in all coordinate systems**. This principle has been applied to states of position (the origin and orientation of axes), as well as to the states of movement of coordinate systems (speed, acceleration). Such states are never defined in an absolute manner, but relatively to one another. For example, there is no absolute movement, in the sense that it can only be defined in a relative way between one body and another.

Galileo introduced explicitly **_velocity_** parameters in the observational referential.

> The special principle of relativity was first explicitly enunciated by Galileo Galilei in 1632 in his **Dialogue Concerning the Two Chief World Systems**, using the metaphor of Galileo's ship:
>
> _"Locked with a friend in the main cabin under the deck of a large ship, and carry with you flies, butterflies, and other small flying animals ... Hang a bottle that empties drop by drop in a large container placed below the same, ... do the boat go with uniform velocity (and no fluctuations in either direction) ... the drops fall into the lower container without departing from the stern, while the boat is advanced while the drops are in the air ... butterflies and flies will continue their flight equally to each side, and will not happen to focus on the stern, as if weary of following the course of the boat ..."_

Then, **Einstein** introduced explicitly **_acceleration_** parameters.

> _An observer in a reference system without communication or visual contact with another reference system cannot determine the speed of a system over another by any experiment._
>
> _But **an accelerated observer moving relative to another observer, yes he can determine the relative value of the acceleration relative to the observer.**_

THE PRINCIPLE OF SCALE RELATIVITY

In a similar way, **Nottale** introduces *scale* parameters explicitly in the observational referential.

Scale relativity proposes in a similar manner to define a scale relative to another one, and not in an absolute way. Only scale ratios have a physical meaning, never an absolute scale, in the same way as there exists no absolute position or velocity, but only position or velocity differences.

The concept of *resolution* **is re-interpreted as the "state of scale" of the system,** in the same way as velocity characterizes the state of movement.

> **The principle of Scale Relativity can thus be formulated as:**
> *"the laws of physics must be such that they apply to coordinate systems whatever their state of scale."*

The main goal of scale relativity is **to find laws which mathematically respect this new principle of relativity.**

Mathematically, this can be expressed through the principle of covariance applied to scales, that is, the invariance of the form of physics equations under transformations of resolutions (dilations and contractions).

> In theoretical physics, *general covariance* (also known as *general invariance*) is the invariance of the form of physical laws under arbitrary differentiable coordinate transformations. The essential idea is that coordinates do not exist a priori in nature, but are only artifices used in describing nature, and hence should play no role in the formulation of fundamental physical laws.
>
> A physical law expressed in a generally covariant fashion takes the same mathematical form in all coordinate systems, and is usually expressed in terms of tensor fields. The classical (non-quantum) theory of electrodynamics is one theory that has such a formulation.
>
> Albert Einstein proposed this principle for his special theory of relativity; however, that theory was limited to space-time coordinate systems related to each other by uniform relative motions only, the so-called "inertial frames." Einstein recognized that the general principle of relativity should also apply to accelerated relative motions.

> *Much of the work on classical unified field theories consisted of attempts to further extend the general theory of relativity to interpret additional physical phenomena, particularly electromagnetism, within the framework of general co-variance, and more specifically as purely geometric objects in the space-time continuum.*

The core idea of scale-relativity is thus to include resolutions explicitly in coordinate systems, thereby integrating measure theory explicitly in the formulation of physical laws: For example, the length of the Brittany coast is explicitly dependent on the resolution at which one measures it.

The relative state of scale is fundamental to know about for any physical description. *For example, if we want to describe the movement and properties of a sphere, we may as well use classical mechanics or quantum mechanics depending on the size of the sphere in question.*

In particular, information on resolution is essential to understand quantum mechanical systems, and in scale relativity, resolutions are included in coordinate systems, so it seems a logical and promising approach to account for quantum phenomena.

SCALE RELATIVITY & GENERAL RELATIVITY

Scientific theories usually do not improve by adding comple-xity, but rather by starting from a more and more simple ba-sis. This fact can be observed throughout the history of science. The reason is that starting from a less constrained basis provides more freedom and therefore allows richer phenomena to be included in the scope of the theory. Therefore, **new theories usually do not contradict the old ones, but widen their domain of validity and include previous knowledge as special cases**. For example, releasing the constraint of rigidity of space led Einstein to derive his theory of general relativity and to understand gravitation. As expec-ted, this theory naturally includes Newton's theory, which is recove-red as a linear approximation under weak fields.

The same type of approach has been followed by Nottale to build the theory of scale relativity. The basis of current theories is a continuous and two-times differentiable space. **Space is by de-finition a continuum, but the assumption of differentiability is**

not supported by any fundamental reason. It is usually assumed only because it is observed that the first two derivatives of position with respect to time are needed to describe motion. **Scale relativity theory is rooted in the idea that the constraint of differentiability can be relaxed and that this allows quantum laws to be derived.**

In terms of geometry, **differentiability means that a curve is sufficiently smooth and can be approximated by a tangent.** Mathematically, two points are placed on this curve and one observes the slope of the straight line joining them as they become closer and closer. **If the curve is smooth enough, this process converges (almost) everywhere and the curve is said to be differentiable.** It is often believed that this property is common in nature.

However, **most natural objects have instead a very rough surface, or contour.** For example the bark of trees and snowflakes have a detailed structure that does not become smoother when the scale is refined. For such curves, the slope of the tangent fluctuates endlessly or diverges. **The derivative is then undefined (almost) everywhere and the curve is said to be non-differentiable.**

Therefore, when **the *assumption* of space differentiability is abandoned, there is an additional degree of freedom that allows the geometry of space to be extremely rough.** The difficulty in this approach is that new mathematical tools are needed to model this geometry because the classical derivative cannot be used.

Nottale found a solution to this problem by using the fact that non-differentiability implies scale dependence and therefore the use of fractal geometry. Scale dependence means that the distances on a non-differentiable curve depend on the scale of observation. It is therefore possible to maintain differential calculus provided that the scale at which derivatives are calculated is given, and that their definition includes no limit. It amounts to saying that non-differentiable curves have a whole set of tangents in one point instead of one, and that **there is a specific tangent at each scale.**

To abandon the hypothesis of differentiability does not mean abandoning differentiability. Instead, **this leads to a more general framework, where *both* differentiable and non-differentiable cases are included.**

Combined with motion relativity, scale relativity by definition thus extends and contains general relativity. As much as general relativity is possible when we drop the hypothesis of euclidian space-time, allowing the possibility of curved space-time, **scale relativity is possible when we abandon the hypothesis of differentiability, allowing the possibility of a fractal space-time.** The objective is then to describe a <u>continuous space-time which is not everywhere differentiable</u>, as it was in general relativity.

Abandoning differentiability doesn't mean abandoning differential equations. The concept of fractal allows work with the non-differentiable case with differential equations. In differential calculus, we can see the concept of limit as a zoom, but in this generalization of differential calculus, one doesn't look only at the limit zooms (zero and infinity) but also everything in between, that is, all possible zooms. In sum, **we can drop the hypothesis of the differentiability of space-time, keeping differential equations, provided that fractal geometries are used.** With them, we can still deal with the non-differentiable case with the tools of differential equations. **<u>This leads to a double differential equation treatment: in space-time and in scale space.</u>**

FRACTAL SPACE-TIME

If Einstein showed that space-time was curved, **Nottale shows that it is not only curved, but also fractal**. Nottale has proven a key theorem which shows that a **space which is continuous and non-differentiable is necessarily fractal.<u>It means that such a space depends on scale.</u>**

Importantly, the theory does not merely describe fractal objects in a given space. Instead, it is *space itself which is fractal.* To understand what a fractal space means requires to study not just fractal curves, but also fractal surfaces, fractal volumes, etc.

Mathematically, a fractal space-time is defined as a non-differentiable generalization of Riemannian geometry. Such a fractal space-time geometry is the natural choice to develop this new principle of relativity, in the same way that curved geometries were needed to develop Einstein's theory of general relativity.

In the same way that general relativistic effects are not felt in a typical human life, **the most radical effects of the fractality of spacetime appear only at the extreme limits of scales: micro scales or at cosmological scales.** This approach therefore proposes to bridge not only the quantum and the classical, but also the classical and the cosmological, with fractal to non-fractal transitions.

MIN & MAX INVARIANT SCALES

A fundamental and elegant result of **Scale Relativity is to propose a minimum and maximum scale in physics, invariant under dilations,** in a very similar way as the speed of light is an upper limit for speed.

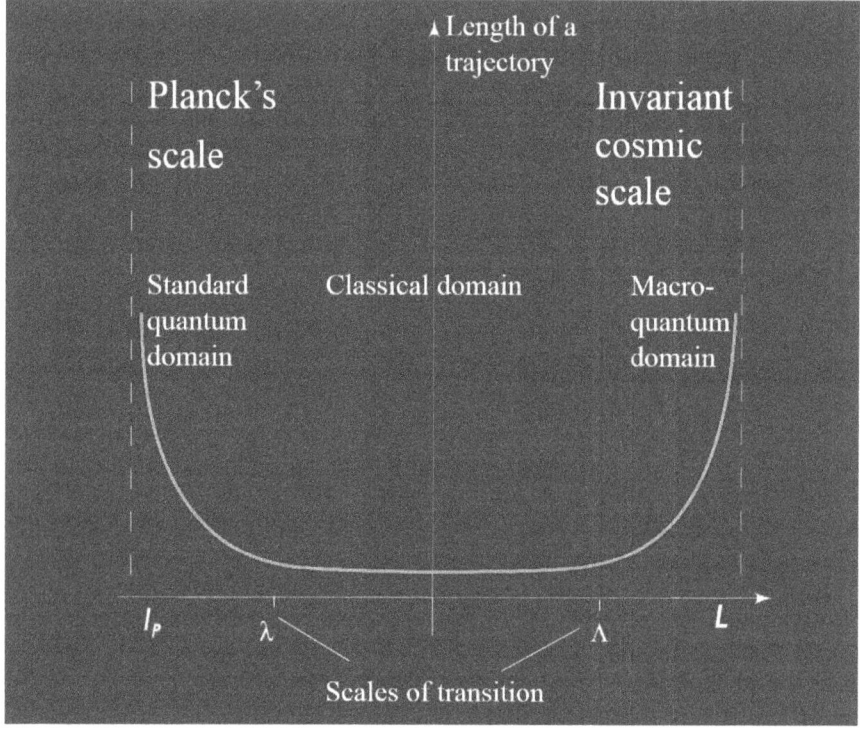

Fig.47: Variation of the fractal space-time geodesics,...

...according to the resolution (scale), in the framework of special scale relativity. The scale symmetry is broken in two transitioning scales λ and Λ (non-absolute), which divide the scale space in three domains: (1) a classical domain, intermediary, where space-time doesn't depend on resolutions because the laws of movement dominate over scale laws; and two asymptotical domains towards (2) very small and (3) very large scales where scale laws dominate over the laws of movement, which makes explicit the fractal structure of space-time. (Wikipedia).

Minimum invariant scale

In Special Relativity, there is an unreachable speed, the speed of light. We can add speeds without end, but they will always be less than the speed of light. The sums of all speeds are limited by the speed of light. Additionally, the composition of two velocities is inferior to the sum of those two speeds.

In Special Scale Relativity, similar unreachable observational scales are proposed, **the Planck length scale (l_P) and the Planck time scale (t_P).** Dilations are limited by l_P and t_P, which means that we can divide spatial or temporal intervals without end, but they will always be superior to Planck's length and time scales.

This is a result of special scale relativity. Similarly, the composition of two scale changes is inferior to the product of these two scales.

Minimum scale l_P = 10 e - 35 m *(Planck scale)*

Maximum invariant scale

The choice of **the maximum scale (noted L)** is less easy to explain, but it mostly consists to identify it with the cosmological constant: **L = 1/(Λ²)**. This is motivated in parts because a dimensional analysis shows that the cosmological constant is the inverse of the square of a length, i.e. a curvature.

As Λ = 10 e -122 , Maximum scale L = 10 e + 244 m

SR VS OTHER RELATIVITY THEORIES

The theory of scale relativity follows a similar construction as the one of the relativity of movement, which took place in three steps: Galilean, Special and General Relativity.

This is not surprising, as in both cases the goal is to find laws satisfying transformation laws including one parameter that is relative: the speed in the case of the relativity of movement; the resolution in the case of the relativity of scales.

Galilean Scale Relativity

Galilean scale relativity involves linear transformations, a constant fractal dimension, self-similarity and scale invariance. This situation is best illustrated with self-similar fractals. Here, the length of geodesics varies constantly with resolution. The fractal dimensions of free particles doesn't change with zooms. These are self-similar curves.

In galilean relativity, recall that the laws of motion are the same in all inertial frames. Galileo famously concluded that "the movement is like nothing". In the case of self-similar fractals, paraphrasing Galileo, one could say that "scaling is like nothing". Indeed, the same patterns occur at different scales, so scaling is not noticeable, it is like nothing.

In the relativity of movement, Galileo's theory is an additive galilean group:

$$X' = X - VT$$
$$T' = T$$

However, if we consider **scale transformations** (dilations and contractions), the **laws are products, and not sums.** This can be seen by the necessity to use units of measurements. Indeed, when we say that an object measures 10 meters, we actually mean the object measures 10 times the definite predetermined length called "meter". The number 10 is actually a scale ratio of two lengths 10/1m, where 10 is the measured quantity, and 1m is the arbitrary defining unit. This is the reason why the group is multiplicative.

Moreover, an arbitrary scale **e** doesn't have any physical meaning in itself (like the number 10), only scale ratios **r=e'/e** have a meaning, in our example, r=10/1. Using the Gell-Mann-Lévy method, we can use a more relevant scale variable, **V=ln (e'/e)**, and then find back an additive group for scale transformations by taking the logarithm, which converts products into sums.

Interestingly, when, in addition to the principle of scale relativity, one adds the principle of relativity of movement, there is a transition of the structure of geodesics at large scales, where trajectories do not depend on the resolution anymore, where trajectories become classical. This explains the shift of behavior from quantum to classical.

Special Scale Relativity

Special Scale Relativity can be seen as a correction of Galilean scale relativity, where Galilean transformations are replaced by Lorentz transformations. Interestingly, the "corrections remain small at "large" scale (i.e. around the Compton scale of particles) and increase when going to smaller length scales (i.e. large energies) in the same way as motion-relativistic corrections increase when going to large speeds".

---

*The **Compton wavelength** is a quantum mechanical property of a particle. It was introduced by Arthur Compton in his explanation of the scattering of photons by electrons (a process known as Compton scattering).*
The Compton wavelength of a particle is equivalent to the wavelength of a photon whose energy is the same as the rest-mass energy of the particle.

*The standard Compton wavelength, **λ**, of a particle is given by*

$$\lambda = \frac{h}{mc}$$

*where **h** is the Planck constant, **m** is the particle's rest mass, and **c** is the speed of light. The significance of this formula is shown in the derivation of the Compton shift formula.*

The CODATA 2010 value for the Compton wavelength of the electron is 2.4×10^{-12} m. Other particles have different Compton wavelengths.

In **Galilean** relativity, it was considered "obvious" that we could add speeds without limit (**$w = u + v$**). This composition laws for speed was not challenged. However, **Poincaré and Einstein** did challenge it with special relativity, setting a maximum speed on movement, the speed of light. Formally, **if v is a velocity, $v + c = c$**. The status of the speed of light in special relativity is a horizon, unreachable, impassable, invariant under changes of movement.

Regarding scale, we are still within a Galilean kind of thinking. Indeed, we assume without justification that **the composition of two dilations is $\rho * \rho = \rho^2$**. Written with logarithms, this equality becomes **$\ln\rho + \ln\rho = 2\ln\rho.$** However, nothing guarantees that this law should hold at quantum or cosmic scales. As a matter of fact, this dilation law is corrected in special scale relativity, and becomes:

lnρ + lnρ = 2 ln ρ / (1 + ln ρ^2).

More generally, in special relativity the composition law for velocities differs from the Galilean approximation and becomes (with the speed of light $c=1$):

$u \oplus v = (u + v) / (1 + u*v)$

Similarly, in special scale relativity, the composition law for dilations differs from our Galilean intuitions and becomes (in a logarithm of base K which includes a possible constant C = ln K, which plays the same role as c):

logρ1 $\oplus$ logρ2 = (logρ1 + logρ2) / (1 + logρ1 *logρ2)

The status of the Planck scale in special scale relativity plays a similar role as the speed of light in special relativity. It is a horizon for small scales, unreachable, impassable, invariant under scale changes, i.e. dilations and contractions. The consequence for special scale relativity is that **applying two times the same contraction ρ to an object, the result is a contraction less strong than contraction ρ x ρ. Formally, if ρ is a contraction, $\rho * l_P = l_P$.**

As noted above, there is also an unreachable, impassable maximum scale, invariant under scale changes, which is the cosmic length L. In particular, **it is invariant under the expansion of the universe.**

General Scale Relativity

In Galilean Scale Relativity, **spacetime was fractal with *constant* fractal dimensions.**

In Special Scale Relativity, **fractal dimensions can vary.**

This **varying fractal dimension means that the laws satisfy a logarithmic version of the Lorentz transformation.** The varying fractal dimension is covariant, in a similar way as proper time is covariant in special relativity.

In **General Scale Relativity, the fractal dimension is not constrained anymore, and can take any value**. In other words, it is the situation where **there is *curvature in scale space***. Einstein's curved space-time becomes a particular case of the more general fractal spacetime.

General Scale Relativity is much more complicated, technical, and less developed than its Galilean and special versions. It involves non-linear laws, scale dynamics and gauge fields. In the case of non self-similarity, changing scales generates a new scale-force or scale-field which needs to be taken into account in a scale dynamics approach. Quantum mechanics then needs to be analyzed in scale space.

Finally, in general scale relativity, we need to take into account both movement *and* scale transformations, where scale variables depend on space-time coordinates. More details about the implications for abelian gauge fields and non-abelian gauge fields can be found in the literature. **Nottale's 2011 book provides the state of the art.**

To sum up, one can see some structural similarities between the re-lativity of movement and the relativity of scales:

Relativity	Variables defining the coordinate system	Variables characterizing the state of the coordinate system
Movement	Space	Speed
	Time	Acceleration
Scale	Length of a fractal	Resolution
	Variable fractal dimension	Scale acceleration

Fig.48: Relativity of movement and relativity of scales.

In both cases, there are two kinds of variables linked to the coordinate systems: variables which define the coordinate system, and variables that characterize the state of the coordinate system. In this analogy, the resolution can be assimilated to a speed; acceleration to a scale acceleration; space to the length of a fractal; and time, to the variable fractal dimension. Table from WIKIPEDIA.

SCALE RELATIVITY VS QUANTUM MECHANICS

The fractality of space-time implies an infinity of virtual geodesics. This remark already means that a fluid mechanics is needed. **The idea to consider a fluid of geodesics in a fractal spacetime is an original proposal from Nottale.**

In scale relativity, quantum mechanical effects appear as effects of fractal structures on the movement. **The fundamental indeterminism and nonlocality of quantum mechanics are deduced from the fractal geometry itself.**

There is an analogy between the interpretation of gravitation in general relativity and quantum effects in scale relativity. Indeed, if gravitation is a manifestation of space-time curvature in general relativity, **quantum effects are manifestations of a fractal spacetime in scale relativity.**

To sum up, **there are two aspects which allows scale relativity to better understand quantum mechanics.** On the one side, *fractal fluctuations themselves* are hypothesized to lead to quantum effects. On the other side, ***non-differentiability*** leads to a local irreversibility of the dynamics and therefore to the use of complex numbers.

Quantum mechanics thus receives not only a new interpretation, but a firm foundation in relativity principles.

SCALE RELATIVITY AND OTHER TOE APPROACHES

It may help to understand scale relativity by comparing it to various other approaches to unifying quantum and classical theories.

String theory

Although string theory and scale relativity start from different assumptions to tackle the issue of reconciling quantum mechanics and relativity theory, the two approaches need not to be opposed. Indeed, **Castro suggested to combine string theory with the principle of scale relativity:**

*"It was emphasized by Nottale in his book that **a full motion plus scale relativity** including all spacetime components, angles and rotations **remains to be constructed**. In particular the **general theory of scale relativity**. Our aim is to show that **string theory provides an important step in that direction** and vice versa: the scale relativity principle must be operating in string theory."*

Quantum gravity

Scale relativity is based on a geometrical approach, and thereby recovers the quantum laws, instead of assuming them. **This distinguishes it from other quantum gravity approaches**. Nottale comments:

*"The main difference is that these **quantum gravity studies assume the quantum laws to be set as fundamental laws**. In such a framework, the fractal geometry of space-time at the Planck scale is a consequence of the quantum nature of physical laws, so that **the fractality and the quantum nature co-exist as two different things**. In the **scale relativity theory**, there are not two things (in analogy with Einstein's general relativity theory in which gravitation is a manifestation of the curvature of space-time): the **quantum laws are considered as manifestations of the fractality and nondifferentiability of space-time**, so that they do not have to be added to the geometric description."*

Loop quantum gravity

They have in common **to start from relativity theory and principles, and to fulfill the condition of background independence**.

El Naschie's E-Infinity theory

El Naschie has developed a similar, yet different fractal space-time theory, because he gives up differentiability *and* continuity. El **Naschie thus uses a "Cantorian" space-time**, and uses mostly number theory. This is to be contrasted with **scale relativity, which keeps the hypothesis of continuity, and thus works preferentially with mathematical analysis and fractals**.

Causal Dynamical Triangulation (CDT)

Through computer simulations of causal dynamical triangulation theory, **a fractal to nonfractal transition was found from quantum scales to larger scales.** This result seems to be compatible with quantum-classical transition deduced in an other way, from the theoretical framework of scale relativity.

Non-commutative geometry

For both scale relativity and non-commutative geometries, **particles are geometric properties of space-time.** The intersection of both theories seems fruitful and still to be explored. In particular, Nottale further generalized this non-commutativity, saying that it "is now at the level of the fractal space-time itself, which therefore fundamentally comes under Connes's noncommutative geometry. Moreover, this noncommutativity might be considered as a key for a future "

CP violation *(CP standing for charge parity) is a violation of the postulated CP-symmetry (or charge conjugation parity symmetry).It plays an important role both in the attempts of cosmology* **to explain the dominance of matter over antimatter in the present Universe.**

Wave–particle duality *is the fact that every elementary particle or quantic entity exhibits the properties of not only particles, but also waves. It addresses the inability of the classical concepts "particle" or "wave" to fully describe the behavior of quantum-scale objects.*

Doubly special relativity (RDS)

Both theories (SR & DRS) **have identified the Planck length as a fundamental minimum scale.** However, as Nottale comments:

"The main difference between the "Doubly-Special-Relativity" approach and the scale relativity one is that we have identified the question of defining an invariant length-scale as coming under a relativity of scales."

SCALE RELATIVITY VS FRACTAL RAINBOW

It is very clear that the proposal of the **"SCALE RELATIVITY"** and his **"FRACTAL SPACE-TIME"** (Nottale) shares many concepts with the proposal offered in this book **Fractal Rainbow** (we might also call the **SCALE-FRACTAL-EMERGNCE proposal**), but they do not share all and in various aspects they both have some conceptual differences.

Both theories share:

- To consider the **(spatial) Scale as a variable** to be considered in determining the physical laws of nature.

- To consider **FRACTAL THEORY** as an option to be considered for establishing the physical laws of nature.

During the development of my second article (Part 2 of this book), when I was evaluating the different current cosmological theories (Strings-branes, Quantum Gravity, Emergency, …) and its "state of the art": what they said about the limits of Our Universe and on what could be beyond them, casually I discovered some references to the Nottale proposals and Theory. This was a pleasant surprise that supported my own ideas (in the first article I already used the concept and name of Scale Relativity of the Universe in the title).

*Although both conceptual proposals are very similar and they assume the same basic concepts, obviously **it is not comparable the degree of mathematical and physical development that had already been done Nottale** (and his team) over the past 30 years (collected in his two books of 1993 and 2011 , see bibliography). What is surprising is how little known is nowadays this theory and how it is still marginalized in the academic domains. **But the day that this theory will be taken seriously by the scientific community (it would be demostrate and verified), I am convinced that it will generate the great leap that requires the current physics since Einstein.***

But also **offer following differences:**

SCALE RELATIVITY establishes the hypothesis that the length and the Planck time scale are limit values, from which there cannot be lower values (they cannot be divided into smaller values). **Planck Scale and Time are considered invariant values.**

While **FRACTAL RAINBOW** doesn't have scale limits and consider that **Planck Scale is only a possible (Brane) Horizon , and it is not invariant.** The Rainbow Fractal accepts that **there may be a limit (Event Horizon) on the KK scale (as 6D Calabi-Yau shapes)**, and that they will shape the basic fabric of space-time of our universe. But also propose that inside them should be other universes (concepts and laws). It is not known that size that could have these forms of Calabi-Yau 6D, but if given the dimension of Planck, **they might be these minimum (invariant) scale proposed by the theory of Nottale.**

So we could say that the **SR**, (even considering the Scale Factor and the Fractal Space-Time) as it is proposed now, it **"only" will reach to model Our Universe**, from the Planck scale to the outer limits of Our Universe, which would be extended (since 10 e +27 m to 10 +244 m).

SCALE RELATIVITY proposal, determine **TWO scale limits of the Universe:**

Min. scale $l_P = 10$ e - 35 m *(Planck length)*

Max. scale   L = 10 e + 244 m *(Universe length)*

Instead, the **FRACTAL RAINBOW theory of the Universe has no limits and end-boundaries**, although could be (Brane) Horizons between different (Brane) scales, and presupposes an infinite Universe, or otherwise, **a very large and wrap-around universe without end.**

SCALE RELATIVITY also proposes a possible unification of physical laws in a ToE and, moreover, it proposes itself as a theory containing the precedent theories SR-GR and QM.

While **FRACTAL RAINBOW reject the option of a single ToE for the Whole Universe** (due to Gödel principle), and it propose a **generation of different EMERGENT laws for each Scale Landscape** (Cosmic, Relativistic, Newtonian, Quantum, Planckian,...), although, **in some way related to each other**. **Different TOEs only will cover wider scale range.**

And could be there, where, possibly, **Fractal Theory could be offered as a underlaying theory** that would help to establish the foundations of these emerging theories, **depending on the space scale values.**

The **SCALE RELATIVITY** has proposed interesting alternatives to concepts that were difficult to understand, mostly for **smallest scales** (wave-particle duality, principles of Heisenberg, Schrödinger and de Broglie, Indeterminism and non-locality, quantum fluctuations, Quantification, ...), based on the assumption of **fractal space-time (non-differential) and its geodesics.**

But SR has also offered solutions and explanations for concepts of the **largest scales** (Dark Matter, Variability G, ...), also based on the assumption of **fractal space-time.**

As already mentioned above, as Einstein proposed the **Curved Spacetime** to explain Gravity, Nottale proposes **Fractal Spacetime** to explain the phenomena of Quantum Mechanics, and the possible "leak" of Gravity and "hiding" Mass in Our Universe.

With its "invariant" scale constraints (MIN-MAX), the **SCALE RELATIVITY is not longer than a theory within a bounded universe (with limits): possibly Our Universe** (between 10 e -35 m to 10 e + 244 m).

While the proposal of the **FRACTAL RAINBOW aims to go beyond the borders (Event Horizons) of Our Universe**, and **assuming new laws, forces and concepts unknown to us today,** beyond these limits.

Although, possibly, **if we combine Scale Relativity with String (Branes) Theory**, it could be obtained a broader scope, considering different branes (universes) contained within one another. Possibly this would be what Nottale described as **The General Theory of Relativity Scale.**

ANNEX 6: GR + QM + TOE THEORIES

This annex only try to give a **reminder overview of the general concepts of Classic, Relativistic and Quantum theories, and the TOE** (Theory of Everything) seeking to unify them.

In physics the term **_theory_ is generally used for a mathematical frame-work**—derived from a small set of basic postulates (usually symmetries, like equality of locations in space or in time, or identity of electrons, etc.)—**which is capable of producing experimental predictions for a given category of physical systems.**

Mechanics is an area of science concerned with the **behavior of physical bodies when subjected to forces or displacements**, and the subsequent effects of the bodies on their environment.

The scientific discipline has its origins in Ancient Greece with the writings of Aristotle and Archimedes. During the early modern period, scientists such as Galileo, Kepler, and Newton, laid the foundation for what is now known as **classical mechanics**. And further Einstein develop the **relativistic mechanics** and collaborate with other physics (Planck, Bohr, Schrodinger, Heisenberg,...) to the **quantum mechanics**.

In physics, **classical mechanics** and **quantum mechanics** are the two major sub-fields of mechanics.

Classical mechanics is concerned with the set of **physical laws describing the motion of bodies under the influence of a system of forces.** The study of the motion of bodies is one of the oldest and largest subjects in science, engineering and technology. It is also widely known as **Newtonian mechanics**.

Classical mechanics describes the motion of macroscopic objects, from projectiles to parts of machinery, as well as astronomical objects, such as spacecraft, planets, stars, and galaxies. Besides this, many specializations within the subject deal with solids, liquids and gases and other specific sub-topics.

Classical mechanics also **provides extremely accurate results** as long as the **domain of study is restricted to large objects** and the **speeds involved do not approach the speed of light.**

When **speeds involved approach the speed of light, or mass involved are very high** (black holes), it becomes necessary to the **Relativistic Mechanics** (Special and General Relativity Theories).

When the **objects being dealt with become sufficiently small,** it becomes necessary to introduce the other major sub-field of mechanics, **Quantum Mechanics**, which **reconciles the macroscopic laws of physics with the atomic nature of matter** and handles the wave–particle duality of atoms and molecules.

When both quantum mechanics and classical mechanics cannot apply, such as **at the quantum level with high speeds, quantum field theory (QFT)** becomes applicable.

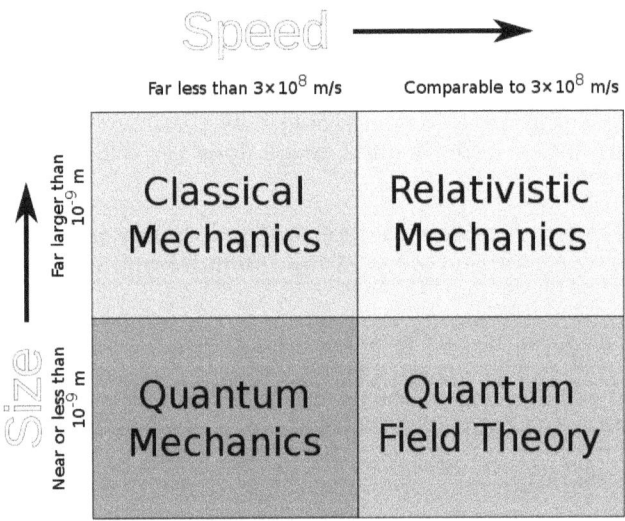

Fig.49: Mechanics Theories

A **theory of everything (ToE)** is a hypothetical single, all-encompassing, **coherent theoretical framework of physics that fully explains and links together all physical aspects of the universe.** Finding a ToE is one of the major unsolved problems in physics. Over the past few centuries, two theoretical frameworks have been developed that, as a whole, most closely resemble a ToE. These two theories upon which all modern physics rests on are general relativity (GR) and quantum field theory (QFT). **GR is a theoretical framework that only focuses on the force of gravity for understanding the universe in regions of both large-scale and high-mass**: stars, galaxies, clusters of galaxies, etc. On the other hand, **QFT is a theoretical framework that only focuses on three non-gravitational forces** for understanding the universe in regions of both **small scale and low mass:** sub-atomic particles, atoms, molecules, etc. QFT successfully implemented the Standard Model and unified the interactions (so-called **Grand Unified Theory**, **GUT**) between the three non-gravitational forces: weak, strong, and electromagnetic force

Some TOEs are currently being evaluated, which aim is to unify these four theories: Classical, Relativistic, Quantum and QFT. The one that has had greater repercussion and more known is the **String Theory (Theory M)** that we have already treated in the annex 4.

CLASSICAL MECHANICS

Broadly, **we can divide the Classical Mechanics (CM) in two main "areas": kinematics and dynamics.**

The **kinematic** states all laws associated with **space (e) and time (t)**: the speed ($v = e / t$) and acceleration ($a = v / t$).

While the **dynamics** also includes the concepts of **mass (m), force (F) and energy (E)**: $F = m a$ y $Ek = 1/2 m v2$ (kinetic energy).

CM perfectly explains the phenomena with movement due to the application of forces or energy on a mass bodies, but on scales of human dimensions (10 e-10 m to 10 e+10 m), and where are not involved very high masses or speeds, in which case it is required **Relativistic Mechanics**.

RELATIVISTIC MECHANICS

The **Theory of Relativity**, usually encompasses two theories by Albert Einstein: **Epecial Relativity (SR) and General Relativity (GR).**

Concepts introduced by the theories of relativity include spacetime as a unified entity of space and time, relativity of simultaneity, kinematic and gravitational time dilation, and length contraction.

As with classical mechanics, the subject can be divided into **"kinematics" (Special Relativity, SR)**; the description of motion by specifying positions, velocities and accelerations, and **"dynamics" (General Relativity, GR)**; a full description by considering energies, momenta, and angular momenta and their conservation laws, and forces acting on particles or exerted by particles.

There is however a subtlety; **what appears to be "moving" and what is "at rest"**—which is termed by "statics" in classical mechanics—**depends on the relative motion of observers who measure in frames of reference.**

Special Relativity (**SR**) is the generally accepted and experimentally well confirmed physical theory regarding the relationship between space and time. In Einstein's original pedagogical treatment, it is based on two postulates: (1) that the **laws of physics are invariant (i.e. identical) in all inertial systems** (non-accelerating frames of reference); and (2) that the **speed of light in a**

vacuum is the same for all observers, regardless of the motion of the light source.

General Relativity (GR), is the current description of gravitation in modern physics. General Relativity **generalizes special relativity and Newton's law of universal gravitation**, providing a unified description of gravity as a geometric property of space and time, or spacetime. In particular, **the curvature of spacetime is directly related to the energy and momentum of whatever matter and radiation are present.**

QUANTUM MECHANICS

Classical electromagnetism or **classical electrodynamics** is a branch of theoretical physics that **studies the interactions between electric charges and currents using an extension of the classical Newtonian model.** The theory provides an excellent description of electromagnetic phenomena whenever the relevant length scales and field strengths are large enough that quantum mechanical effects are negligible. **For small distances and low field strengths, such interactions are better described by quantum electrodynamics.**

Quantum Mechanics (QM) including quantum field theory, is a **fundamental branch of physics concerned with processes involving**, for example, **atoms and photons**. In such processes, said to be quantized, the action has been observed to be only in integer multiples of the Planck constant, a physical quantity that is exceedingly, indeed perhaps ultimately, small. This is utterly inexplicable in classical physics.

QUANTUM FIELD THEORY

Quantum Field Theory (QFT) is a theoretical **framework for constructing quantum mechanical models of subatomic particles in particle physics and quasiparticles in condensed matter physics**. A QFT treats particles as excited states of an underlying physical field, so these are called field quanta.

In quantum field theory, quantum mechanical interactions between particles are described by interaction terms between the corresponding underlying fields.

Quantum electrodynamics (QED) has one electron field and one photon field; **quantum chromodynamics (QCD)** has one field for each type of quark; and, in condensed matter, there is an atomic displacement field that gives rise to phonon particles. Edward Witten describes **QFT as "by far" the most difficult theory in modern physics.**

Quantum Electrodynamics (QED) is the relativistic quantum field theory of electrodynamics. In essence, it **describes how light and matter interact and is** the first theory **where** full agreement between **quantum mechanics and special relativity is achieved**. QED mathematically describes all phenomena involving electrically charged particles interacting by means of exchange of photons and represents the quantum counterpart of classical electromagnetism giving a complete account of matter and light interaction.

QED (jointly the Pauli principle) explains how works the molecule and atom structures. The interactions between electrons and protons due to the electromagnetic charges.

> The **Pauli exclusion principle** is the quantum mechanical principle that states that two identical fermions (e.i. electrón) cannot occupy the same quantum state simultaneously. In the case of electrons, it can be stated as follows: it is impossible for two electrons of a poly-electron atom to have the same residing in the same orbital.

Quantum Chromodynamics (QCD) is the theory of strong interactions, a fundamental force **describing the interactions between quarks and gluons** which make up hadrons such as the proton, neutron and pion. The QCD analog of electric charge is a property called _color_. Gluons are the force carrier of the theory, like photons are for the electromagnetic force in quantum electrodynamics. The theory is an important part of the Standard Model of particle physics. A large body of experimental evidence for QCD has been gathered over the years.

QCD enjoys two peculiar properties:

- **Confinement**, which means that the force between quarks does not diminish as they are separated. Strong Force is stronger as the quarks separate between them.
- **Asymptotic freedom**, which means that in very high-energy reactions, quarks and gluons interact very weakly creating a quark–gluon plasma.

QCD explains how **quarks make hadrons (protons & neutrons)**, and how **bosons establish the interaction forces between different fermions** (G, EM, S & W).

STANDARD MODEL

The **Standard Model** is the theoretical **framework describing all the currently known elementary particles**.

This model contains:
- Six flavors of **quarks** (q), named up (u), down (d), strange (s), charm (c), bottom (b), and top (t).
- Six types of **leptons**, known as _flavours_, forming three _generations_. The first generation is the _electronic leptons_, comprising the electron (e⁻) and electron neutrino (v_e); the second is the _muonic leptons_, comprising the muon (μ⁻) and muon neutrino (v_μ); and the third is the _tauonic leptons_, comprising the tau (τ⁻) and the tau neutrino (v_τ).
- And five types of **bosons** (g) gluon, (V) photon, (Z-W) weak and (H) Higgs.

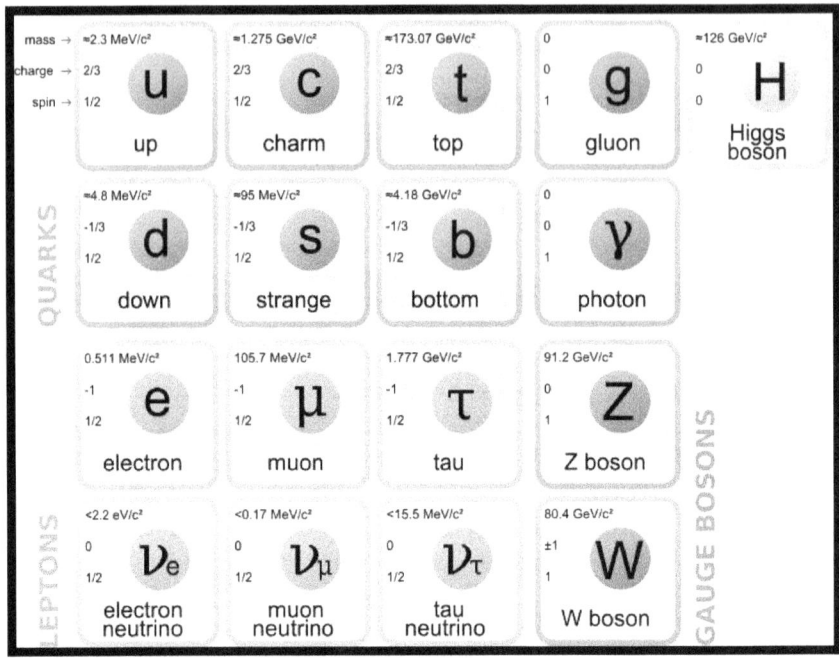

Fig.50: Standard Model particles: Quarks, Leptons & Bosons

TOE

Through years of research, physicists have experimentally confirmed with tre-
mendous accuracy virtually every prediction made by these two theories when
in their appropriate domains of applicability. In accordance with their findings,
scientists also learned that **GR and QFT, as they are currently formulated,
are mutually incompatible** - they cannot both be right. Since the usual domains of applicability of GR and QFT are so different, most situations require
that only one of the two theories be used. As it turns out, **this incompatibility
between GR and QFT is only an apparent issue in regions of extremely
small-scale and high-mass**, such as those that exist within a black hole or
during the beginning stages of the universe (i.e., the moment immediately following the Big Bang). To resolve this conflict, a theoretical framework revealing
a deeper underlying reality, **unifying gravity with the other three interactions, must be discovered to harmoniously integrate the realms of GR
and QFT** into a seamless whole: a single theory that, in principle, is capable of
describing all phenomena. In pursuit of this goal, **quantum gravity** has re-
cently become an area of active research.

Over the past few decades, a single explanatory framework, called **"string
theory"**, has emerged that may turn out to be the ultimate theory of the universe. Many physicists believe that, **at the beginning of the universe** (up to

10^{-43} seconds after the Big Bang), **the four fundamental forces were once a single fundamental force**. Unlike most (if not all) other theories, string theory may be on its way to successfully incorporating each of the four fundamental forces into a unified whole. According to string theory, **every particle in the universe**, at its most microscopic level (Planck length), **consists of varying combinations of vibrating strings (or strands)** with preferred patterns of vibration. String theory claims that it is through these specific oscillatory patterns of strings that a particle of unique mass and force charge is created (that is to say, the electron is a type of string that vibrates one way, while the up-quark is a type of string vibrating another way, and so forth).

CM PRINCIPLES:

The **Newton Theory** of gravitation **works perfectly with energies and velocities we found in everyday life** (in Our reference Scale).

But **Newton Theory fails** to the ends: **For very high speeds** (Near to the speed of light) **and very high energy and mass**. There, the **Relativity Theory (of Einstein) unseat the Newtonian one.**

GR-SR PRINCIPLES:

The basic postulates of **Special Relativity (SR)** of Einstein says (*"inertial system" means constant speed*):
• The notions of space and time can be treated as a **single structure of spacetime**.
• The **laws of physics are the same in all "inertial systems".**
• The **speed of light is the same in any "inertial system".**

The basic postulates of **General Relativity (GR)** of Einstein says:
• The **Equivalence Principle** (GR): The effects of acceleration are indistinguishable from those of gravity.
• **Gravity is due to deformation of spacetime.**

QM PRINCIPLES

The strange **new concepts found in the QM** are: quantification, the wave function, the wave-particle duality and the uncertainty principle.

• **Quantification:** The proceses has been observed to be only in integer multiples of the Planck constant. This is utterly inexplicable in classical physics.
• **Wave Function:** Any particle has a wave function associate that determine the probability of its position.
• **Wave-Particle duality:** Every elementary particle or quantic entity exhibits the properties of not only particles, but also waves.
• **Uncertainty Principle:** Is a fundamental limit to the precision with which certain pairs of physical properties of a particle, known as complementary variables, such as position x and momentum p, can be known simultaneously.

12. EPILOGUE

This book, besides presenting a **"new" approach (framework) of the universe** (including the concept of scale, and the different Scale Landscapes that may exist), it tried to **include some studies, reflections and opinions that are currently in scientific field that could be related to its main idea or proposal** (Emergence, Fractal, Scale Relativity, Hologram, String-branes,...). **But also how this proposal could affect to some common concepts** (Energy, Matter, Time, Vacuum,...) **as also to other unusual concepts** (Dark Matter and Energy, Quantum Fluctuations, Uncertain Principle, Wave-Particle Duality,...)

When **we show all these concepts on the same "framework" (Scale Landscapes),** we hope **it could help-us to link ideas and proposals**, which currently are being treated separately. And it can provide an **essential link between them**, offering the possibility that different scientists (physics, mathematics, cosmological,...) can bring new ideas and proposals that **help all us to better understand the whole universe and its laws.**

It is remarkable the great (physical) scientific advances that have been made throughout the history of humanity, mainly during the last 100-500 years (Pythagoras, Aristotle, Newton, Maxwell, Einstein, Susskind,...).

But it is also true that (considering that current mainstream accept that known energy and matter are not more than 5% of all the forecast energy and matter of Our Known Universe) **we are still long way from understanding the whole universe in all its amplitude and complexity.**

Concepts as common as energy (matter), time and space, are being challenged and redefined. There are still many important voids / gaps / inconsistencies in our knowledge of the universe. And it just makes their study so interesting and intriguing.

As we have seen, at present **we only know and understand 5% of the matter / energy of (Our) Universe**, while the rest is known as Dark Energy and Matter. But if we consider the scalar proposal of this book (FRACTAL RAINBOW), it would be more appropriate to say that **we know very well the concepts and laws that govern our Reference Scale** (from 10 e-10 to 10 e +10 m, the Newtonian Landscape) , while **we become more difficult to understand and parameterize the concepts and laws of the other Scalar Landscapes as they move away from ours.**

To date, we have been able to understand and model our spectrum scale (from 10 e-20 m to 10 e+20 m). But, from now on, it is becoming more complex to understand the spectra that are beyond. Therefor, **we will need to break with pre-established ideas and concepts, and be able to accept new schemes.**

In this book the following questions still **remain to be resolved in the (near?) Future:**

- **How should we include the _Scale Factor_ to improve the _Dynamic_ laws of Newton and Einstein: _F = m.a.S_ (being S = Scale factor).**

- **How should _Scale Factor_ help to understand the _Plank Landscape_ and smaller scale landscapes?**

- **How should we use _Fractal Theory "tool"_ to parametrize the nD fractal space-time, and to get a _wider ToE_ ?**

These three issues could be some of the main cosmological problems to be solved during the next future (next decade ?).

MOND Theories (and similar like TeVeS,...) have done good proposals on the first issue, while the **Theory of Scale Relativity** of Nottale, after 30 years of development, has also done to date breakthroughs and explanations over the last two issues.

This book proposes a new approach to take into account for the study of these scales more distant to the Our:

- For **large-scale sizes**, the possibility that **new forces/ interactions arise due to the clustering of stars / galaxies** that cause new effects. Just as the Force of Gravity emerges as large numbers of atoms-molecules come together.

- And **for small-scale sizes**, also could **exist forces/ interactions that we do not know** and that do not affect Our Reference Scale.

---

*The **SCALE RELATIVITY** has proposed interesting alternatives to concepts that were difficult to understand, mostly for **smallest scales** (wave-particle duality, principles of Heisenberg, Schrödinger and de Broglie, Indeterminism and non-locality, quantum fluctuations, Quantification, ...), based on the assumption of **fractal space-time (non-differential) and its geodesics.***

*But SR has also offered solutions and explanations for concepts of the **largest scales** (Dark Matter, Variability G, ...), also based on the assumption of **fractal space-time.***

*As Einstein proposed the **Curved Spacetime** to explain Gravity, Nottale proposes **Fractal Spacetime** to explain the phenomena of Quantum Mechanics, and the possible "leak" of Gravity and "hiding" Mass in Our Universe.*

The proposal of the Fractal, Emergent and Scale Relativity of the present book, if one day it is definitively accepted (demonstrated and verified) by the "scientific community", **could completely revolutionize the cosmological physics,** assuming a jump as important as was Theory of Relativity (Special and General) a century ago. And those scientists, who can establish a consistent General Theory in this basis **(Nottale?), could be considered as deserved substitutes of Einstein, Newton and Aristotle.**

Possibly, **if we combine Scale Relativity with String (Branes) Theory**, it could be obtained a broader scope. And this would be what Nottale described as **The General Theory of Relativity Scale.**

In the same way that, to date, these breakthrough discoveries have been due to great minds (working independently in laboratories or offices), from now, **it will require teamwork** to address the challenges that we still have. And this implies a new way of approaching science, **by applying the multidisciplinary teamwork methodologies and framework (*the "synergy effect"*).**

I hope that this book can spark ideas and fields of study for the next generation of scientists ("post-string" and "post-QG"): **New Scale-Fractal-Emergent generations.**

Since the purpose of this book is to present mainly a conceptual proposal, but it is not including any physical theoretical demonstration or any experimental verification, we could say that, in addition to its own informative character about the current "state of the art" of cosmology, **its content could be better classified as Scientific Philosophy, than as Cosmological Physics.**

"Strictly speaking, there is no science that is not based on assumptions. The idea that it is not, it is unthinkable, it is a paralogical thought. Always must exist a previous philosophy, a "belief ", so that the science ,from it, remove a direction, a meaning, a limit, a method, a right to exist.".
Friedrich Nietzsche, "On the Genealogy of Morality"

DEFINITIONS

Whole Universe: Everything that exists, although we can not see, or even imagine, and therefore we may be unable to model and parameterize.

Our (Known) Universe : Our Pocket Universe (within the Whole Universe) that was generated in Our Big-Bang and in which we exist and that we are able to understand and model.

Observable Universe: That part of our universe that (due to the limitation of the speed of light) we are able to detect or observe.

Pocket Universe: Those other universes (like Our Universe) that could exist beyond Our Universe (on the Whole Universe).

Scale Landscape: Scalar Spectrum of the Whole Universe that could have their own emergent concepts and laws.

Cosmic Horizon: The event horizon (boundary) could be between Our Universe (Our 4D-Brane), and beyond of Our Universe (the Bulk or Cosmic Landscape).*(Approx. > 10 e +30 m).*

Observable Horizon: The event horizon (boundary) of the Observable Universe (that part of Our Universe that, due to the limitation of the speed of light, we are able to detect y observe). *(Approx. 10 e+27 m)*

Planck Horizon: The event horizon (edge or boundary) where (1. If edge) end Our Universe spectra on the lower scales, or (2.- If boundary) where it make the change between Our Universe spectra (Our 4D-Brane) with the Sub-Planck Landscape (possibly within 6D space Calabi-Yau forms-branes-universes). **(< 10 e - 35 m)**

Brane Horizon: The event horizon (boundary or edge) between two or several branes, possibly of different (space) dimensions (3D-2D, 3D-4D, 3D-6D,...). Also could be named **Dimensional Horizons.**

BIBLIOGRAPHY

Brian Greene:
- *"The Hidden Reality: Parallel Universes and the Deep Laws of the Cosmos"*, 2011.
- *"The Fabric of the Cosmos: Space, Time, and the Texture of Reality"*, 2005.
- *"The Elegant Universe: Superstrings, Hidden Dimensions, and the Quest for the Ultimate Theory"*, 1999.

Leonard Susskind:
- *"The Cosmic Landscape"*, 2005

John D. Barrow:
- *"The Constants of Nature"*, 2002

Lee Smolin:
- *"The Trouble With Physics: The Rise of String Theory, the Fall of a Science, and What Comes Next"*, 2006

Stephen Hawking:
- *"The Grand Design"* with Leonard Mlodinow , 2010.
- *"A Brief History of Time"*, 1988.

Lisa Randall:
- *"Warped Passages: Unraveling the Universe's Hidden Dimensions. 2005*

Robert B. Laughlin:
- *"Different Universe: Reinventing Physics from the Bottom Down"*, 2006

Robert Temple:
- *"The Crystal Sun"*, 2000

Sir Arthur Eddington:
- *"Fundamental Theory"*, 1953

Laurent Nottale:
- *"Scale Relativity And Fractal Space-Time: A New Approach to Unifying Relativity and Quantum Mechanics"*. 2011
- *"Fractal space-time and microphysics:towards a theory of scale re-lativity."* 1999.

Roger Penrose:
- *"Cycles of Time"*. 2011

Martin Bojowald:
- *"Zurück vor den Urknall"*. 2010 ("Before the Big-Bang")

Amanda Gefter:
- *"Trespassing on Einstein's Lawn"*. 2015

Max Tegmark:
- *"Our Mathematic Universe"*. 2014

ARTICLES

Previous Author´s Articles:
"The ¨Matryoshka-verses¨: The scale relativity of the Universe" (David Piñana, October 2012). http://matryoshka-dimension.blogspot.com.es.
"The Scale Landscapes (Relativity) of the Universe" (David Piñana, October 2015).

John Maldacena: "The Illusion of Gravity" (Scientific American, January-2006): http://www.scientificamerican.com/article/the-illusion-of-gravity/

Frank Wilczek: "What´space" (2009). http://web.mit.edu/physics/news/physicsatmit/physicsatmit_09_whatisspace_wilczek.pdf

Stephen Hawking: "Gödel and the end of physics", 2002. http://www.hawking.org.uk/godel-and-the-end-of-physics.html

Nathan Seiberg: "Emergent Spacetime", 2006 (http://arxiv.org/find/hep-th/1/au:+Seiberg_N/0/1/0/all/0/1)

Fu Yuhua, Fu Anjie, Zhao Ge: *"Fifteen Kinds of Waves Caused by Four Fundamental Forces" (Beijing Relativity Theory Research Federation) (http://gsjournal.net/Science-Journals/Research%20Papers-Relativity%20Theory/Download/4346)*

Vincent J. Martinez & Bernard J.T. Jones: *"Why the universe is not a (simple) fractal (but yes a multi-fractal)". (1990)*: http://adsabs.harvard.edu/abs/1990MNRAS.242..517M

"Violation of Heisenberg's Measurement-Disturbance Relationship by Weak Measurements" **(Lee A. Rozema, Ardavan Darabi, Dylan H. Mahler, Alex Hayat, Yasaman Soudagar, and Aephraim M. Steinberg** Phys. Rev. Lett. 109, 100404 – Published 6 September 2012; Erratum Phys. Rev. Lett. 109, 189902 (2012). http://journals.aps.org/prl/abstract/10.1103/PhysRevLett.109.189902

FQXi ESSAY CONTEST: 2006 FQXi The Nature of Time (http://fqxi.org/community/essay/winners/2008.1)
• The Nature of Time by **Julian Barbour.**
• Does Time Exist in Quantum Gravity? by **Claus Kiefer.**

"A Fractal Universe?" (**Robert L. Oldershaw**, 2002, http://www3.amherst.edu/~rloldershaw/NOF.HTM)

Laurent Nottale:
The Theory of Scale Relativity (1991):
Scale relativity and fractal space-time: theory and applications (2009)

GRATEFULNESSES

Leonard Susskind (Theoretical Physics at Stanford University). who was one of the few first level expert that answer my first article (Oct.2012) and give a short opinion about.

Eduard Salvador (Physicist Prof. Cosmology UPC with various works and articles on Dark Matter), who after reading the book V 1.0, told me that the general proposal could be correct, but without being demonstrated and verified, it could be Considered better as Philosophy of Science (Philosophy of Cosmology).

Francesc Fayos (Physical Prof. Master of Cosmology of the UPC) to which he was the first who got the Second Article in August 2015, and helped me to present it (without success) to referees and to the ARXIV WEB page.

Juan Jose Curto (Physicist Dtr. "Observatori del Ebre", Roquetes-Tortosa), with whom I comment the different versions of the articles and the final book.

Salvador Tarragó (Architect Prof. UPC) who after the purchase of the book called me to comment and give his opinion always so useful.

Antonio Coso (Computer science), who after acquiring the book V 1.0 sent me a Errata Sheet for his correction, as well as his opinion about the book as a Physicist amateur.

Antonio Dalmau (Ex-Editor), who gave me his opinion as editor about the book as well as editing support.

Jaume Escriva (Ing. Ind. UPC), with whom I shared and discussed my first article of October 2012, and who propose the Fractal Theory as a possible solution to model a wider range of dimensional scales spectra (Scalar Landscapes).

Enric Galera (Dipl. English Language and Humanistic Science, UCB), with whom we had long philosophical discussions about the content of this article, and who reviewed its English (on both articles Oct.2012 & 2015).

Forums WEB (and its members) that I used to discuss this proposal, and also I learned over mainstream theories and studies on this
field, although I was constantly banned by making no mainstream proposals and questions:

The Science Forum (http://www.thescienceforum.com). Members: Strange, Markus Hanke, Implicate Order, guymillion,...

CosmoQuest Forum (https://cosmoquest.org). Members: Shaula, Strange, Ken G, John Mendenhall, ShinAce, Reality Check,...

Foro 100cia (http://e-ciencia.com/opinion/foros/). Members: Teaius, Javiucm,...

WIKIPEDIA:

I appreciate the great help that I got by using Wikipedia to obtain information on the topics covered in this book (and previous articles), as well as, by the reproduction of certain texts and figures (mainly in the annexes and for general definitions and historical info).

Therefore, due to this contribution, it has been agreed to fund WIKIPEDIA with part of the royalties gotten through the sale of this book.

LIST OF FIGURES

Fig.1: References of scale of Our Universe   24
Fig.2: Images:Brain Cell and Universe Galaxies   25
Fig.3: Levels of scales of the Global Universe   30
Fig.4-5: Possible 4D shapes for Our Universe   33
Fig.6: Electromagnetic Radiation Spectrum 38
Fig.7: Gravitational waves   41
Fig.8: Possible collisions on the WMAP 44
Fig.9: Positive Scale Landscapes   50
Fig.10: Negative Scale Landscapes   51
Fig.11: The Landscapes of the Universe   55
Fig.12: Cosmic Landscape (3D representation)   58
Fig.13: Matter-energy contents of Our Universe   61
Fig.14: CDT Planck Landscape texture 68
Fig.15: 6D Calabi-Yau shape   71
Fig.16: Relation between Energy & Length scales. 73
Fig.17: DSR Diagram [c= F(E)] 76
Fig.18: Point of View from different Landscapes   84
Fig.19: Emergent Interaction Fields   86
Fig.20: Interaction field unification   87
Fig.21: Scope of action of the Interaction Field   88
Fig.22: Universe Event Horizons   95
Fig.23: Black Hole Horizon   96
Fig.24: Observable Universe Horizon 98
Fig.25: CMB (Cosmic Microwave Background)   99
Fig.26: Hologram theory 101
Fig.27: Escher´s circle as Fractal & Hologram sample   106
Fig.28: Cosmological Constant varied since the BB 114
Fig.29: Game of Life (John Horton Conway , 1970) 119
Fig.30: Comparison rotation curves (galaxy M33) 124
Fig.31: Image of the Bullet Cluster   130
Fig.32: MOND (TeVeS) & DarkMatter predictions.   131
Fig.33: Variation of G (Gravity Constant) with Scale   133
Fig.34: Example of emergence in a physical system: 137
Fig.35: Exact self-similarity   148
Fig.36: Quasi self-similarity   148
Fig.37: Statical self-similarity   149
Fig.38: Multifractal scaling   149
Fig.39: Measuring the Coast of Britain 150
Fig.40: Matter distribution in a cubic section of the Universe.   153
Fig.41: 2D Branes showing Gravity   159
Fig.42: KK branes 160
Fig.43: Calabi-Yau Branes   161
Fig.44: Calabi-Yau Structure   161
Fig.45: Multi D-Braneworlds in the Bulk   162
Fig.46: Gravity Interactions between D-Braneworlds   164
Fig.47: Variation of the fractal space-time geodesics,...   172
Fig.48: Relativity of movement and relativity of scales.   177
Fig.49: Mechanics Theories   186
Fig.50: Standard Model particles: Quarks, Leptons & Bosons   190

BACK COVER

> "A book that could revolutionize the future of Cosmological Physics: Aristotle, Newton, Einstein,..."

The author presents a vision of the Universe from a totally different point of view, and in a disclosure way, and very easy to understand.

It is a journey from the smallest (the dimension of Planck) to the largest (Our Universe boundary). And he also shows, in a clear way, which may be beyond these limits.

The new proposals on Scale Landscape and Scale Relativity raised in this book could be a breakthrough in the current "state of the art" of the cosmology, showing a new outlook for a better understanding of the Universe.

This book will change our view about some common concepts (Energy, Matter, Time, Vacuum ...) and also about other "unusual" concepts (Dark Matter and Energy, Quantum Fluctuations, Uncertainty Principle, Wave-Particle Duality,....), based on recent studies and theories (Emergence, Fractal, Scale Relativity, Holography, String-Branes, Quantum Gravity, ...).

Required reading for both: physics-cosmological experts, to explore an innovative proposal, as well as general public, that just would like to learn more about the Universe from a different and original point of view.

www.ingramcontent.com/pod-product-compliance
Lightning Source LLC
Chambersburg PA
CBHW051458170526
45166CB00001B/295